MACMILLAN MSM SECONDARY
MATHEMATICS

LIVHBEC

BOOK 4ˣ

Jim Miller

Deputy Head, Redewood School,
Newcastle upon Tyne and Assistant Chief
Examiner in GCSE Mathematics

Graham Newman

Head of Mathematics, Prestwich High School, Bury
and Chief Examiner in GCSE Mathematics

Macmillan Secondary Mathematics Series Editor:
Dr Charles Plumpton

M
MACMILLAN

First Published 1990

Published by
MACMILLAN EDUCATION LTD
Houndmills, Basingstoke, Hampshire RG21 2XS
and London
Companies and representatives
throughout the world

Cover design by Plum Books, Southampton

Cover picture © Daily Telegraph Colour Library

Printed in Hong Kong

British Library Cataloguing in Publication Data
Miller, Jim
Macmillan secondary mathematics.
Bk. 4X
1. Mathematics
I. Title   II. Newman, Graham
510
ISBN 0–333–34609–2

Contents

Preface .... **v**

Acknowledgements .... **vi**

1. Money (1) **1**

Percentages; Changing percentages into fractions and decimals; Finding percentages of quantities; Increase/decrease by a percentage; Changing fractions and decimals into percentages; One quantity expressed as a percentage of another; Percentage profit and loss

2. Money (2) **16**

Simple interest; Compound interest; Car insurance; Life assurance; Loans

3. Algebra (1) **28**

Introduction; More than one letter; Subtraction; Multiplication; Division; Indices; Brackets; Factorising

4. Algebra (2) **49**

Substituting numbers and letters for words; Evaluating formulas; Equations; Solving equations: addition, subtraction, fractions, brackets, letters occurring on both sides, subtraction of the unknown value; Transforming formulas

Revision exercises: Chapters 1–4 .... **68**

5. Communication: numbers (1) **75**

Order of operations; Repetitive operations; Positive indices; Standard form notation; Negative indices and standard form; Negative indices and integers; Rounding; Significant figures; Reciprocals; Squares; Square roots; Pythagoras; Using Pythagoras' theorem

6. Communication: numbers (2) **100**

The number line; Inequalities; Addition of directed numbers; Subtraction of directed numbers; Zeros; Multiplication of directed numbers; Division of directed numbers; Problem-solving; Sets; Comparing sets; Venn diagrams; Combining sets

7. Communication: graphics (1) **122**

Co-ordinates; Negative co-ordinates; Straight-line graphs; Using a table; Non-linear equations; Gradient; Solving simultaneous equations graphically

8. Communication: graphics (2) **142**

Bisecting a line; Bisecting an angle; Drawing angles; Loci

9. Communication: geometry **152**

Congruence; Similarity; Angles between parallel lines; Polygons; Angles in a polygon; Regular polygons; Angle in a semicircle; Angle between tangent and radius

10. Communication: statistics **171**

Histograms; Line graphs; Pie charts; Pictograms; Frequency distributions; Tally marks; Mode and median; Quartiles; Grouped data – mode, median; Scatter graphs; Line of best fit

Revision exercises: Chapters 5–10 210

11. Measuring 233

Fractions: additions, subtractions, multiplication, division; Decimals: applications

12. Measuring: shapes and solids 246

Area and perimeter; Area of a parallelogram; Circumference and area of a circle;
Combined shapes; Volume; The cylinder; Problems with volume; Edge, face and vertex

13. Probability 264

Experimental probability; Combined probabilities; Selection without replacement

14. Trigonometry 275

Right-angled triangles; Ratio of sides; Sine; Using calculators; Using sine; Finding
the angle; Cosine; Using cosine; Tangent; Using tangent; Summary; Terminology

Revision exercises: Chapters 11–14 294

Preface

This book forms part of the Macmillan Secondary Mathematics course which has been written to correspond to the National Criteria for GCSE Mathematics examinations, and to the Mathematics National Curriculum assessment and testing framework. In particular the X-stream books are for the use of pupils who intend going on to the lower or intermediate levels of the GCSE Mathematics examination.

Book 4X further extends the skills which have already been used in Book 3X. Each chapter contains worked examples, and sets of graded exercises. In addition there are many mathematics investigations which can be used to supplement skills learnt, and offer extensions to the exercises. Three revision exercises are included which will also provide for additional practice if needed.

In terms of the National Curriculum, the work covered in Book 4X continues the work in Book 3X up to Levels 6 and 7. Several topics in Level 8 are also included which are required by GCSE syllabuses at the intermediate level. Book 4X covers both the profile components, attributing weight corresponding to the weightings given in the attainment targets. Many examples used are in the context of 'using and applying mathematics', and in realistic contexts of everyday life and at work, within today's multicultural society.

1. Money (1)
Percentages

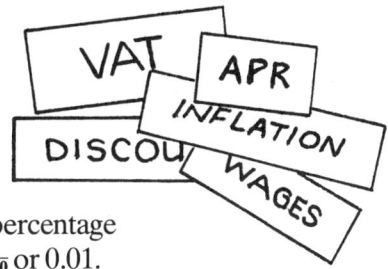

Percentages were first introduced in Book 3X, Chapter 3. A percentage was defined as a part of 100, that is, 1 per cent is the same as $\frac{1}{100}$ or 0.01. To represent a percentage as a fraction and a decimal is useful in that it reminds us that a percentage is indeed a part of a whole. Calculators make it easier for us to work out percentages.

Changing percentages into fractions and decimals

A percentage is actually a fraction of 100, so 1% is $\frac{1}{100}$, and 7% is $\frac{7}{100}$.

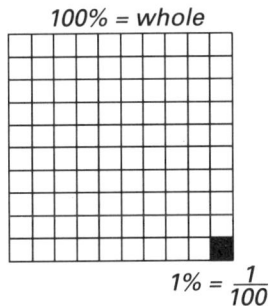

100% = whole

$1\% = \frac{1}{100}$

We already know how to change these fractions into decimals:

$\frac{1}{100}$ means $1 \div 100 = 0.01$ $\frac{7}{100}$ means $7 \div 100 = 0.07$

Example 1
Write 85% (*a*) as a fraction, (*b*) as a decimal.

(*a*) $85\% = \frac{85}{100} = \frac{17}{20}$

(*b*) $\frac{85}{100}$ means $85 \div 100 = 0.85$

EXERCISE 1.1

Write the following percentages as fractions, simplifying your answer where possible.

1 50%   **2** 75%   **3** 10%   **4** 17%   **5** 20%   **6** 35%   **7** 95%

8 5%   **9** 49%   **10** 23%   **11** 3%   **12** 150%   **13** 225%   **14** 62%

15 80%

Write the following percentages as decimals.

16 50%   **17** 25%   **18** 15%   **19** 19%   **20** 40%   **21** 55%   **22** 100%

23 65%   **24** 90%   **25** $12\frac{1}{2}$%   **26** $7\frac{1}{2}$%   **27** $18\frac{1}{2}$%   **28** 99%   **29** 33%

30 21%

EXERCISE 1.2

Write the following percentages as (*a*) a fraction, (*b*) a decimal.

1 30%   **2** 60%   **3** 45%   **4** 85%   **5** 1%   **6** 98%   **7** 24%

8 27%   **9** 34%   **10** 51%

Investigation A
Try to write $12\frac{1}{2}$% in terms of a fraction. What happens? Investigate and find a method which would work for all such percentages. Try $33\frac{1}{3}$%, $24\frac{1}{4}$%, etc.

Investigation B
Try to change $33\frac{1}{3}$% into a decimal. What happens? Can you find any other percentage which would be difficult to write as a decimal? What do all these types of percentage have in common?

Finding percentages of quantities

As the whole quantity is 100% we can find what 1% represents first, then the percentage required.

Example 2
Find 15% of £3.60.

$$£3.60 \div 100 \times 15 = £0.54$$

     ↑   ↑
  for 1% for 15%

Example 3
Find 35% of £2.90.

$$£2.90 \div 100 \times 35 = £1.015$$
$$= £1.02 \quad \text{to the nearest penny}$$

EXERCISE 1.3

Calculate the following, rounding your answers to the nearest penny where necessary.

1 10% of £54   **2** 5% of £5   **3** 26% of £100   **4** 12% of £2.50

5 8% of 25p   **6** 14% of £3   **7** $12\frac{1}{2}$% of £8   **8** $15\frac{1}{2}$% of £6

9 $4\frac{1}{4}$% of £24   **10** 2% of £2   **11** 9% of 90p   **12** 15% of 10p

13 $12\frac{1}{2}$% of £3   **14** 5% of £4.70   **15** $4\frac{1}{2}$% of £37.56

16 A saleswoman is paid 8% commission on her sales of £760. How much does she receive?

17 A silicon wafer of 100 chips has 5% which are faulty. (*a*) How many are faulty? (*b*) How many are not faulty?

18 Of 6000 runners in a marathon, 12% did not finish the race. (*a*) How many people failed to finish the race? (*b*) How many did finish?

FINISH

19 What is the VAT at 15% on a television set costing £280?

20 A workforce of 86 people is invited to share in the £1898.88 monthly net profit of a company. They can share $12\frac{1}{2}$% of the profit between them. How much does each person receive?

Increase/decrease by a percentage

Example 4

(*a*) Increase £85.50 by 10%. (*b*) Decrease £93.00 by 5%.

(*a*) 10% of £85.50 is £85.50 ÷ 100 × 10 = £8.55.
So the increased amount is £85.50 + £8.55 = £94.05.
(*b*) 5% of £93.00 is £93.00 ÷ 100 × 5 = £4.65.
So the decreased amount is £93.00 − £4.65 = £88.35.

EXERCISE 1.4

1 Increase (*a*) £5 by 10% (*b*) £23 by 5% (*c*) £100 by $7\frac{1}{2}$% (*d*) £150 by 3%.

2 Decrease (*a*) £4 by 10% (*b*) £48 by 25% (*c*) £200 by $8\frac{1}{4}$% (*d*) £300 by $2\frac{1}{2}$%.

3 Calculate the total cost of the following after the addition of VAT at 15%: (*a*) a radio for £18 (*b*) a coat for £23 (*c*) a case for £15.20.

4 Find the new price of (*a*) a £23 car tyre with a discount of 10%, (*b*) a £53 electric fan heater with a discount of 5%, (*c*) a £380 TV set reduced by 32%.

5 House prices have increased by $10\frac{1}{2}$% in the last 12 months. What would be the value now of a house bought 12 months ago for £52 000?

6 A factory workforce, originally of 520 people, has been reduced by 35% over a period of time. How many people are there in the workforce now?

7 The value of a company on the stock exchange falls by 16%. If the initial value was £3 million, what will be the value of the company now?

8 Find the total cost of a hi-fi unit advertised at £445 + 15% VAT.

EXERCISE 1.5

1 The total sales of a pop group's first record were $1\frac{1}{2}$ million. Their second record released sold 38% more copies than the first. How many of the second record were sold?

2 A company producing heaters finds that 4% are usually faulty. Out of a batch of 500 how many will (*a*) be faulty, (*b*) work perfectly?

   The production is increased and the next batch produced has 10% more heaters. (*c*) How many heaters are in this next batch? (*d*) How many of these will be faulty?

3 The graph shows the percentages of each class of vehicle recorded out of a sample of 150. Write down the actual number of vehicles recorded in each category of (*a*) cars (*b*) lorries (*c*) vans (*d*) buses.

   There were some other vehicles which could not be included in these four categories. (*e*) What percentage could be classified as 'other' vehicles? (*f*) How many of these other vehicles were there in the sample?

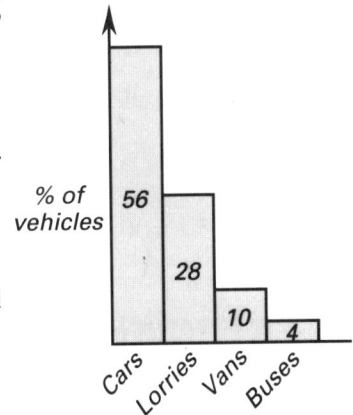

4 An electrical component is made up of three circuit boards A, B and C. Samples tested indicated that, of the final product, 7% will have a failure of one circuit board, 3% a failure of two circuit boards, and just $\frac{1}{2}$% a failure of all three circuit boards. The components are produced in batches of 1000.

   (*a*) What percentage of the electrical components will have no faults?

   (*b*) Out of a single batch of 1000 produced, how many components will have a failure in (i) one (ii) two (iii) three circuit boards?

   (*c*) Out of each batch, 40% are sold overseas. How many components are sent to overseas customers?

   (*d*) What percentage of those components sold overseas will be returned with a failure of all three circuit boards?

Changing fractions and decimals into percentages

Whereas a fraction is normally part of a whole, a percentage is part of a hundred. When we want to write a fraction or a decimal as a percentage we therefore need to make them 100 times greater, by multiplying by 100.

Example 5

Write (a) $\frac{4}{5}$ as a percentage (b) 0.155 as a percentage.

(a) $\frac{4}{5} \times 100 = \frac{400}{5} = 400 \div 5 = 80\%$

(b) $0.155 \times 100 = 15.5\% = 15\frac{1}{2}\%$

EXERCISE 1.6

Write the following fractions as percentages.

1 $\frac{1}{2}$ **2** $\frac{1}{4}$ **3** $\frac{1}{10}$ **4** $\frac{1}{20}$ **5** $\frac{7}{10}$ **6** $\frac{9}{20}$

7 $\frac{1}{5}$ **8** $\frac{4}{5}$ **9** $\frac{23}{50}$ **10** $\frac{17}{100}$ **11** $\frac{4}{25}$ **12** $\frac{49}{50}$

13 $\frac{7}{25}$ **14** $\frac{1}{100}$ **15** $1\frac{1}{4}$ **16** $2\frac{1}{2}$ **17** $\frac{1}{8}$ **18** $\frac{3}{8}$

19 $\frac{7}{8}$ **20** $\frac{1}{16}$ **21** $\frac{5}{16}$ **22** $\frac{15}{16}$ **23** $\frac{11}{16}$ **24** $1\frac{7}{20}$

25 $4\frac{9}{16}$

EXERCISE 1.7

Write the following decimals as percentages.

1 0.2   **2** 0.4   **3** 0.32   **4** 0.57   **5** 0.41   **6** 0.95

7 1.5   **8** 1.25   **9** 2.55   **10** 0.01   **11** 0.045   **12** 0.445

13 0.975   **14** 0.4125   **15** 0.1575

16 Write $\frac{19}{20}$ as (a) a percentage (b) a decimal.

17 (a) What fraction of the diagram is shaded?

(b) What percentage of the diagram is shaded?

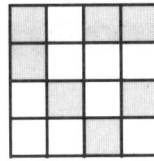

18 (a) What fraction of the diagram is shaded?

(b) What percentage of the diagram is shaded?

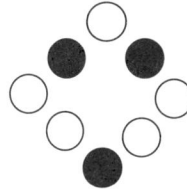

19 In a survey, $\frac{3}{10}$ of the class preferred cheese and onion crisps. Find the percentages of children who preferred the flavours (a) cheese and onion (b) plain (c) salt and vinegar (d) beef.

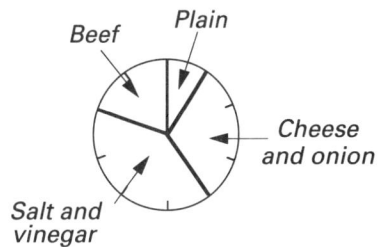

New sign

Old sign

20 An old type of road sign would show a gradient of 1:4 or $\frac{1}{4}$. On new road signs this is shown as a percentage, as indicated in the diagram. What percentages will be shown on the new signs which replace these?

(a)

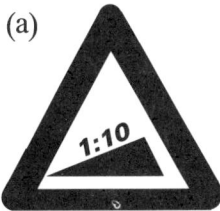

(b)

(c)

(d)

21 $\frac{2}{5}$ of the children on a home economics course were boys.

(a) What percentage were boys? (b) What percentage were girls?

One quantity expressed as a percentage of another

Percentages are frequently used to express one quantity as a percentage of another. This also provides a comparison as to their relative sizes.

Example 6

In a class there are 18 boys and 12 girls. (*a*) What percentage of the class are boys? (*b*) What percentage are girls?

(*a*) Let us first ask ourselves what *fraction* of the class are boys.

Total number in class $= 18 + 12 = 30$

Fraction of boys in class $= \dfrac{18}{30}$

We already know how to change this into a percentage:

$$\frac{18}{30} \times 100 = \frac{1800}{30} = 1800 \div 30 = 60\%$$

(*b*) Since the whole class represents 100%, the percentage of girls is

$$100\% - 60\% = 40\%$$

Example 7

Express 15p as a percentage of £2.

We must remember to use the same units, so £2 is written as 200p.

15p as a fraction of £2 is therefore written as $\dfrac{15}{200}$. As a percentage this is

$$\frac{15}{200} \times 100 = \frac{1500}{200} = 7.5\% = 7\tfrac{1}{2}\%$$

EXERCISE 1.8

1 During September rainfall was recorded on 12 days. What percentage of days of the month (*a*) had rainfall recorded, (*b*) were free from rain?

2 A group of 40 men and women in a factory have jointly won an amount of money on the football pools. What percentage of the winnings will each person receive?

3 In a year group of 180 pupils, 117 are girls. (*a*) What percentage are girls? (*b*) What percentage are boys?

4 A football team scored 51 goals in a season, whereas the reserve team scored 34. What percentage of the total goals scored does this represent for (*a*) the main team, (*b*) the reserve team?

5 After putting £1.20 into a gambling machine a man received 15p back. What is this as a percentage of his money gambled?

6 Over a year a woman spends £75 on football pool coupons, and is then awarded a dividend of £12.60. Write this as a percentage of the money she has spent.

7 A pupil gains the following scores in a series of tests: (*a*) 48 out of 60 (*b*) 12 out of 20 (*c*) 42 out of 48. Write these as percentage results.

8 100 g of butter is to be cut from a pack of 250 g. What percentage is this of the whole pack?

9 In an examination, 18 out of 50 pupils failed. (*a*) What percentage failed? (*b*) What percentage passed?

10 Of 24 learners taking driving tests, 16 passed. What percentage failed?

EXERCISE 1.9

1 A woman earning £120 per month was awarded an increase of £7.80. What percentage is this of her original wage?

2 A newspaper boy was given an extra 4 papers to deliver with his normal 32. What is the percentage increase above the normal delivery?

3 There are 200 children in a year group. On one particular day, 28 arrive late. On this day, what percentage of the year group was (*a*) on time, (*b*) late?

4 Martin is given £2.50 pocket money on Sunday. He spends 65p on Monday and 80p on Tuesday. What percentage of his pocket money does he spend on (*a*) Monday, (*b*) Tuesday? (*c*) By Wednesday, what percentage of his pocket money is left?

5 There were 200 applicants for four jobs at a hotel. Ten people were offered interviews. What percentage of the applicants were offered (*a*) an interview, (*b*) a job? (*c*) What percentage of those offered interviews were eventually given a job?

6 There are 40 oranges in a box, of which 2 are bad. What percentage are (*a*) bad, (*b*) fit to eat?

7 A survey of 50 women was carried out to find what dress sizes were most popular. The results were:

Dress size	8	10	12	14
No. of women	9	16	20	5

Find the percentage of the sample wearing each dress size.

8 A school team played 64 football matches during a season; 24 were drawn games, but 36 were won. What percentage of the matches were (*a*) drawn, (*b*) won, (*c*) lost?

9 Of 400 trains arriving at a station, 64 were late, and 12 were early. What percentage of the trains were (*a*) late, (*b*) early, (*c*) on time?

10 A record is kept of the services required by customers at a hair salon:

Haircut 24	Cut and blow dry 21	Tinting 5
Perm 8	Shampoo and set 25	

Calculate the percentage of customers requiring each service, to the nearest whole number.

Investigation C

A class in a school has 15 boys and 15 girls. What percentage are boys? A boy leaves, to be replaced by a girl. How does this affect the percentage of boys? What happens to the percentage if more boys are replaced by girls? What happens to the percentage of girls in the class at the same time?

Investigation D

Have you ever considered how much time is spent on each of the subjects you study? Make a list of all the subjects you do at school, and the time spent on each. Calculate the total time spent in lessons during a week, and find the percentage of time spent on each of your subjects.

Investigation E

Percentages are used by newspapers and television programmes to describe change, and to inform people. Over a period of time watch out for references to percentages on television and in newspaper or magazine articles. Make a list of what the article or story was about, and how percentages were used. What were percentages most commonly used to describe?

Percentage profit and loss

Many businesses, stores and services operate in order to make a **profit**, that is to earn more than it costs them to operate, so as to gain money. Occasionally this might not be the case, and if money is lost then it is termed a **loss**. Within the retail trade, business is mainly concerned with buying goods, then selling them to someone else at a higher price to make a profit. The percentage profit is the profit expressed as a percentage of the original price, or *cost price*. The percentage loss is the loss expressed as a percentage of the cost price.

Example 8

A radio is bought for £12 and sold for £15. What is the percentage profit?

£12 is the cost price, and £15 is the selling price.

Actual profit $= £15 - £12 = £3$

Profit as a fraction of the cost price is $\dfrac{£3}{£12}$

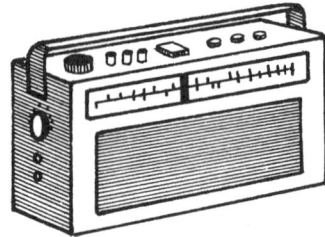

The percentage profit is therefore

$$\dfrac{3}{12} \times 100 = 300 \div 12 = 25\%$$

From this example we can see that:

$$\text{percentage profit} = \dfrac{\text{profit}}{\text{cost price}} \times 100$$

$$\text{percentage loss} = \dfrac{\text{loss}}{\text{cost price}} \times 100$$

EXERCISE 1.10

For the following questions find the percentage profit.

1 Cost price £20, profit £5. **2** Cost price £25, profit £1.

3 Cost price £260, profit £6.50. **4** Cost price £10, profit 50p.

5 Cost price £125, profit £2.50.

For the following questions find the percentage loss.

6 Cost price £80, loss £2. **7** Cost price £120, loss £1.80.

8 Cost price £30, loss £3. **9** Cost price £3.20, loss 8p.

10 Cost price £500, loss 75p.

EXERCISE **1.11**

For each problem find the percentage profit or percentage loss, stating clearly whether the percentage is a profit or a loss.

1 Cost price £12, selling price £18.

2 Cost price £25, selling price £29.

3 Cost price £100, selling price £90.

4 Cost price £125, selling price £106.25.

5 Cost price £45, selling price £54.45.

6 Cost price £230, selling price £181.70.

7 Cost price £3.50, selling price £2.87.

8 Cost price £1030, selling price £1158.75.

9 A watch bought for £39.50 and sold for £35.55.

10 A pendant purchased for £25 and sold for £27.

11 Gas cookers bought for £300, which have a resale price of £246.

12 A new car cost £7500 and was sold a year later for £5550.

13 A sewing machine with a cost price of £120 is sold for £140.40.

14 Wheelchairs cost £265 from the manufacturers, and are then sold for £312.70.

15 A pair of matching curtains are bought for £36, and are then sold for £30.24.

16 A solitaire diamond ring was bought for £680 and sold for £965.60.

17 A washing machine originally purchased for £369 has a resale price of £339.48.

18 A quilt cover was bought for £27 and sold on the market at a price of £33.48.

19 New wardrobes cost £125, but when sold again attracted a price of £85.

20 A television is bought wholesale for £250, and sold retail for £290.

Example 9

A recliner chair is bought for £180, and is to be sold to make a $12\frac{1}{2}\%$ profit. At what price should it be sold?

Profit that must be made is

$$£180 \times \frac{12.5}{100} = £22.50$$

So selling price is £180 + £22.50 = £202.50.

EXERCISE 1.12

Calculate the selling price needed to give the stated profit or loss in each case.

1 Cost price £20, profit 11%.   **2** Cost price £75, profit 6%.

3 Cost price £125, loss 15%.   **4** Cost price £27.50, loss 30%.

5 Cost price 56p, profit 25%.   **6** Cost price £2500, profit $2\frac{1}{2}\%$.

7 Cost price £4000, loss $\frac{1}{2}\%$.   **8** Cost price £12, loss 4%.

9 Cost price £52, profit 23%.   **10** Cost price £7.50, profit 20%.

EXERCISE 1.13

In each case, calculate the selling price.

1 A delivery of potatoes costs £850, but is sold at a loss of 18%.

2 A market trader makes a profit of 23% on a dress bought for £5.

3 Shares cost a dealer £2550, but then have to be sold with a loss of $12\frac{1}{2}\%$.

4 A car is bought for £2600, and is then sold to make a profit of 5%.

5 A man bought a large jar of sweets for £9, and made a profit of 7% on their sale.

6 A lawnmower originally cost £60, but is then resold at a loss of 15%.

7 A clock, bought for £16.50, is sold at a loss of 30%.

8 An antique teapot originally bought for £5.50 is sold at a profit of 350%.

9 Profits of 28% are made by a shop which buys bicycles for £120.

10 A computer bought for £250 is sold as outdated equipment at a loss of 45%.

Investigation F

A shopkeeper loses 6% by selling a damaged cooker for £188. Find the cost price he paid originally for the cooker. (Note that we *cannot* simply find 6% of the £188 and add it on, since the 6% taken off originally was 6% of the cost price, not the selling price.)

2. Money (2)

Simple interest

Many of us have accounts with banks, building societies, or the Post Office, and are familiar with interest we receive for putting money in an account. The money we receive is determined by the rate of interest we are given on the account, and can vary from time to time, but is always expressed as a percentage of the amount of money we have in the account.

POST OFFICE

A calculation of **simple interest** is a method used which gives us an approximate guide to the interest we might receive on money left in an account over a period of time. In practice, with the aid of computers, our interest is actually worked out day by day, but we shall use simple interest as an approximate guide, assuming we withdraw the interest from the account once it has been credited to it.

Example 1

An amount of £150 is left in a savings account and earns a rate of interest of 9%. Find out how much interest is awarded after (*a*) 1 year (*b*) 3 years.

(*a*) We first need to find 9% of £150

$$£150 \div 100 \times 9 = £13.50 \quad \text{interest per year.}$$

(*b*) As £13.50 interest is awarded each year, over three years we will have been given three times as much.

$$£13.50 \times 3 = £40.50 \quad \text{interest over 3 years.}$$

EXERCISE 2.1

For questions 1 – 10, find the simple interest.

1 £350 for 1 year in an account which earns $8\frac{1}{2}$% interest.

2 £155 for 2 years in an account which earns 8% interest.

3 £430 for 4 years in an account which earns 11% interest.

4 £625 for 5 years in an account which earns 10% interest.

5 £80 for 3 years in an account which earns 9% interest.

6 £200 for 4 years in an account which earns $9\frac{1}{2}$% interest.

7 £310 for $2\frac{1}{2}$ years in an account which earns 10% interest.

8 £700 for $3\frac{1}{2}$ years in an account which earns 8% interest.

9 £440 for 6 years in an account which earns 9% interest.

10 £275 for 4 years in an account which earns 11% interest.

11 A lady keeps £315 for 3 years in an account earning 9% interest. Find the total interest gained after this period of time.

12 £250 is left in a savings bank for 2 years. Find the interest earned at a rate of 11%.

13 £750 is deposited in an account, and is left there for 5 years at a rate of $10\frac{1}{2}$%. Find the total interest gained in this period of time.

14 £130 earns an interest rate of 9%. What will be the total interest earned after 3 years?

15 An amount of £98 is left in a savings account for 4 years at a rate of interest of 10%. Calculate the total interest earned at the end of the 4 years.

Example 2

Trevor has £112.50 in an account which earns $9\frac{1}{2}$% interest. How much interest will Trevor receive after (*a*) 1 year (*b*) 2 years (*c*) 3 months?

(*a*) £112.50 ÷ 100 × 9.5 = £10.68$\underline{75}$ = £10.69
    This amount of money does not make sense until we round off the 75 to give the answer.

(*b*) £10.69 × 2 = £21.38

(*c*) We need to write 3 months as a fraction of a year. Since there are 12 months in a year we can write 3 months = $\frac{1}{4}$ year

   Interest over 3 months = £10.69 ÷ 4 = £2.6725
                                        = £2.67   to the nearest penny.

EXERCISE 2.2

Find the simple interest on each of the following.

1 £125 for 5 years at $7\frac{1}{2}$%.   **2** £160 for 3 years at $9\frac{1}{2}$%.

3 £360 for 2 years at 8%.   **4** £410 for $2\frac{1}{2}$ years at 11%.

5 £625 for $1\frac{1}{2}$ years at 10%.   **6** £414 for 6 months at 10%.

7 £132 for $1\frac{1}{4}$ years at 9%.   **8** £514 for 3 months at $9\frac{1}{2}$%.

9 £237 for $2\frac{1}{4}$ years at 8%.   **10** £101 for $1\frac{3}{4}$ years at $8\frac{1}{2}$%.

11 Danielle invests an amount of £500 in an account for $3\frac{1}{2}$ years. What interest will she have received after this period of time at a rate of $12\frac{1}{2}$%?

12 Alan has a savings account which earns 9% interest. Estimate the interest he will have gained if £123 is left in the account for $2\frac{1}{4}$ years.

13 Bobby puts his first £10 in the 'Kiddies Saver' account which earns 10% interest. How much interest will he get after one month?

14 Shaheen has £126 to put in a deposit account at a rate of 8% interest. How much interest will have been credited to her account after $2\frac{1}{2}$ years?

15 The £312 in Peter's account has been earning $9\frac{1}{2}$% interest over $1\frac{1}{4}$ years. How much interest has Peter gained in this time?

Compound interest

Simple interest is an approximation to what we might receive, and we assumed the interest was withdrawn from the account once we had received it. In practice this is not the case, as we usually leave any interest in the account to accumulate, or add up. Any further interest we then receive will be based on the amount originally in the account *plus* any interest added to date. **Compound interest** means the interest is made up of several different amounts of interest. How will this affect the interest we receive? We will work through the figures from Example 1 again, this time adding the interest on each year.

Example 3

An amount of £150 is left in a savings account and earns a rate of interest of 9%. Find out how much interest is awarded after (*a*) 1 year (*b*) 3 years.

(*a*) We first need to find 9% of £150

$\quad$ £150 ÷ 100 × 9 = £13.50 $\quad$ interest in the first year.

(*b*) We will now have £150 + £13.50 = £163.50 in the account.
To find the interest for the second year:

$\quad$ £163.50 ÷ 100 × 9 = £14.715 = £14.72 to the nearest penny.

At the end of the second year we will have £163.50 + £14.72 = £178.22 in the account.
To find the interest for the third year:

$\quad$ £178.22 ÷ 100 × 9 = £16.0398 = £16.04 to the nearest penny.

At the end of the third year we will therefore have £178.22 + £16.04 = £194.26 in the account.
$\quad$ The total interest we have earned = £13.50 + £14.72 + £16.04 = £44.26.
$\quad$ Alternatively:

Interest = final amount − original amount = £194.26 − £150 = £44.26

$\quad$ Using simple interest in Example 1 we had an answer of £40.50 for the interest over 3 years. Compound interest is clearly a better estimate of what we would receive.

EXERCISE 2.3

Find the total compound interest for questions 1 – 10.

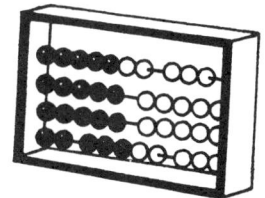

1 £100 for 2 years at 9%. $\qquad$ **2** £150 for 3 years at 10%.

3 £350 for 2 years at 8%. $\qquad$ **4** £250 for 2 years at $9\frac{1}{2}$%.

5 £80 for 2 years at $10\frac{1}{2}$%. $\qquad$ **6** £120 for 3 years at 9%.

7 £125 for 2 years at $9\frac{1}{2}$%. $\qquad$ **8** £75 for 3 years at 8%.

9 £236 for 3 years at 11%. $\qquad$ **10** £309 for 2 years at $10\frac{1}{2}$%.

11 A man has £120 invested in an account earning 9% interest. After 1 year the interest rate rises to $9\frac{1}{2}$% and he keeps his investment in the account for a further year. What total interest will he have received after the 2-year period?

12 A woman invests £245 in an account which earns 8% interest in the first year, then 9% interest for the following two years. What is the total interest she will have received?

Investigation A

Take some of the questions you have answered in Exercises 2.1 and 2.2 and recalculate the interest using compound interest methods to find out how much more interest you estimate you would have received.

Investigation B

Start with any amount of money (to make things easy you could start with £100). Using a rate of interest of 20% and the method of compound interest find out how long it will be before the amount of money in the account is doubled. Repeat the procedure with a smaller interest rate, or a different amount of money to find the effect the changes might have on the length of time.

Car insurance

Car insurance is required for a car by law. It protects other people whenever there is an accident, but it can also protect the driver and the car. There are three types of car insurance:

> *Third party:* protects other people (third party) in case of an accident in addition to the driver (first party) and any passengers (second party).
> *Third party, fire and theft:* as for third party, but also provides insurance in case the car is stolen or involved in a fire.
> *Comprehensive:* an all-purpose insurance which also protects the car itself even if the accident is the fault of the driver.

The gross premium paid is determined by factors such as the type and size of car, the age of the driver, the area in which your car is kept, and for what the vehicle is used. From this gross premium you are allowed certain deductions, the most common being 'No-Claims Discount' (NCD), which varies depending on how many years you have driven without making a claim on your insurance.

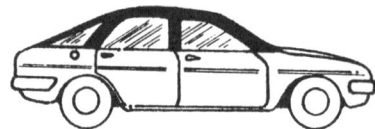

No Claims Discount

Years of no claim	0	1	2	3+
% discount	30	40	50	60

Example 4

Alan's car insurance quotation is for £235. What will it be after he has deducted a no claims discount for 1 year?

1 year's no claims discount is 40%.

40% of £235 is £235 ÷ 100 × 40 = £94

Net premium is £235 − £94 = £141.

EXERCISE 2.4

1 Debbie has been quoted a gross car insurance premium of £260. She is entitled to a 40% no claims discount, having already driven for one year without having an accident. What will she actually be asked to pay?

2 Colin's insurance premium is £325 before deductions. He has driven for two years without having an accident. How much will he have to pay?

3 Adrian has driven for five years without an accident. If the gross premium quoted is £295, how much will be his net car insurance after deduction of NCD?

4 What will be the car insurance on a car after a deduction of 40% NCD from a gross premium of £362?

5 An insurance firm has asked Amanda to pay £220 net for her car insurance policy. To reduce the bill further she agrees to be named as the only driver for a further discount of $7\frac{1}{2}$%. By how much will she reduce the premium?

6 John's gross insurance premium last year was £380, with a 50% NCD. He is told his gross premium will remain unchanged this year. What will be his new net premium after deduction of his NCD?

7 Derek's insurance is £360 before deduction of 60% NCD. He has an accident, and the following year his NCD is reduced to 30%. How much more does it cost him to insure his car?

8 An insurance firm offers Josie a net premium of £190 for the insurance on her motorbike. She would like to reduce it further by becoming the only driver, as this will give her an additional discount of 5%. How much will she then have to pay?

9 National Car Insurance has sent an insurance renewal notice to Gareth. He is entitled to a 60% NCD on a gross premium of £420. (*a*) What is his new premium? (*b*) He has his net premium reduced by a further 10% by agreeing to a voluntary excess. What is his new net premium?

10 Vivek's car insurance is £300 gross per year. He is hoping to claim a 60% NCD for the next three years, but after an accident can only expect to receive a NCD of 30%, 40% and 50% at most over the next three years. Assuming his gross premium stays at £300, what will be the total amount he has lost in insurance premiums through having the accident?

Investigation C

Some car insurance companies allow various percentage deductions from the gross premium quoted. For example, with a gross premium of £418 they could allow:

Mature driver 5%; Insured and spouse only drivers $7\frac{1}{2}$%; Damage excess 5%; No Claims Discount 60%

(*a*) Find the total percentage deductions and hence the net premium.
   This is only an approximate way of finding the net premium, as the insurance companies deduct the percentage discounts *in turn*, starting with the largest, as we did in calculating compound interest.
(*b*) Deduct 60% from £418, then deduct $7\frac{1}{2}$% from your answer, and so on. How does your final answer compare with that from part (*a*)?
(*c*) Try different combinations of percentages with various gross premium amounts.

Life assurance

Life assurance is a type of long-term savings plan which has developed out of concern for relatives. If a man who is the only wage-earner in the family were to die, his family might find it difficult to manage financially. If, however, the wage-earner has a life assurance policy this would give the family an income on which to live. It is called 'assurance' rather than insurance as the person insured does not benefit directly from it, though some policies can be used to provide an additional pension after retirement, or a lump sum on retirement.

EXERCISE **2.5**

The table shows the amount paid out on death as long as the premiums are paid until retirement.
What is the initial sum assured for the following people?

1 A man of age 35 paying £6 per month.

2 A woman of age 20 paying £9 per month.

3 A woman of age 51 paying £6 per month.

4 A man of age 49 paying £6 per month.

5 A woman of age 42 paying £9 per month.

6 A man of age 54 paying £6 per month.

7 A man of age 30 paying £9 per month.

Initial sum assured for an initial monthly premium of:		**£6.00**	**£9.00**
YOUR AGE NOW		**SUM ASSURED £**	
Male	*Female*		
18 - 30	18 - 34	52000	82000
31	35	49367	77848
32	36	44827	70689
33	37	40414	63730
34	38	36448	57476
35	39	32773	51680
36	40	29433	46415
37	41	26351	41554
38	42	23636	37272
39	43	21138	33333
40	44	18886	29782
41	45	16883	26623
42	46	15057	23745
43	47	13448	21206
44	48	12018	18952
45	49	10729	16918
46	50	9605	15147
47	51	8666	13666
48	52	7831	12349
49	53	7058	11131
50	54	6341	10000
51	55	5689	8971
52	---	5104	8049
53	---	4540	7159
54	---	4045	6379
55	---	3609	5691

Investigation D

Divide into groups and discuss the following questions:

1 Why do elderly people receive less benefit (less assurance) than younger people?

2 Females four years older than males receive the same benefit. Why is this?

3 The table is for non-smokers. Why should there be different life assurance policies for smokers and non-smokers?

4 How would you expect the sum assured figures to be different for smokers?

Loans

We sometimes need to obtain loans for many different things: for furniture, carpets, cars, or to do home improvements. In taking out the loan we have to be sure that we can meet the repayments. Some loans cost more than others: the government insists each cash loan shows the Annual Percentage Rate (APR). The APR of a loan is based on the average interest paid on a loan, expressed as an annual percentage rate. This is a guide as to how expensive the loan will be to repay. The higher the APR the more we will have to repay in interest.

EXERCISE 2.6

Use the Home Improvement Loan Repayment Table to answer these questions.

Home Improvement Loan Repayment Table at APR 20%

A Amount of Loan £	**500**	**1,000**	**3,000**	**5,000**
Repayment Term 12 months				
B Monthly Repayment £	45.93	91.86	275.57	459.28
C Total Payable £ *	551.16	1,102.32	3,306.84	5,511.36
Repayment Term 18 months				
B Monthly Repayment £	31.99	63.98	191.95	319.91
C Total Payable £ *	575.82	1,151.64	3,455.10	5,758.38
Repayment Term 24 months				
B Monthly Repayment £	25.05	50.10	150.31	250.52
C Total Payable £ *	601.20	1,202.40	3,607.44	6,012.48
Repayment Term 30 months				
B Monthly Repayment £	20.91	41.82	125.47	209.11
C Total Payable £ *	627.30	1,254.60	3,764.10	6,273.30
Repayment Term 36 months				
B Monthly Repayment £	19.17	36.34	109.02	181.69
C Total Payable £ *	654.12	1,308.24	3,924.72	6,540.84

1 Find the monthly repayments on a loan of £1500 over 36 months.

2 Find the monthly repayments on a loan of £3500 over 24 months.

3 Find the total payable on a loan of £2500 over 12 months.

4 Find the total payable on a loan of £4000 over 30 months.

5 Malcolm needs to borrow £2000 to pay for a new car and can afford repayments up to a maximum of £100 per month. Describe the types of loan which are available to him.

6 Maralyn has already made four payments totalling £200.40 on a loan. How much has she yet to pay?

Amount of Loan Required	72 Months		
	Total Interest	Total Repayable	Monthly Payment
£	£	£	£
2000	1289.68	3289.68	45.69
2500	1611.92	4111.92	57.11
3000	1934.88	4934.88	68.54
3500	2257.12	5757.12	79.96
4000	2579.36	6579.36	91.38
4500	2902.32	7402.32	102.81
5000	3224.56	8224.56	114.23
5500	3546.80	9046.80	125.65
6000	3869.76	9869.76	137.08
6500	4192.00	10692.00	148.50
7000	4514.96	11514.96	159.93
7500	4837.20	12337.20	171.35

Use the 72-month loan repayment table to answer these questions.

7 Find the monthly repayments on a loan of £3000.

8 What is the total repayable on a loan of £7000?

9 What might be the total interest on a loan of £9500?

10 Joan can afford a maximum of £125 per month in repayments, and wants to take out a loan to buy a new car. What is the most she can borrow over the 72-month period?

EXERCISE **2.7**

Use the Low Calendar Monthly Repayment Table to answer the questions.

LET OUR FIGURES DO THE TALKING

APR	GENUINE LOW MONTHLY REPAYMENTS				
	EXAMPLES	36M	60M	90M	120M
14.5%	£20,000 £16,000	- - - - - -	462.05 369.64	356.38 284.80	306.70 245.36
15.3%	£10,000 £5,000	171.87 343.73	234.77 117.39	182.32 91.16	157.69 78.84
15.7%	£4,500 £3,000	155.27 103.52	106.28 70.85	82.96 55.31	71.70 47.80
16.7%	£2,500 £2,000 £1,500	87.40 69.92 52.44	60.27 48.22 36.16	47.29 37.83 28.38	41.26 33.01 24.76

1 Find the monthly repayments on a loan of £3000 over 5 years.

2 Find the total payable on a loan of £2500 over $7\frac{1}{2}$ years.

3 What does it cost in interest in settling a loan of £5000 over 10 years?

4 How much would the repayments be for a loan of £10 000 over 3 years?

5 What could be the total repayments on a loan of £3750 over 5 years?

Use the monthly repayment table to answer the questions.

Monthly Repayments

Amount	3 yrs	5 yrs	7 $\frac{1}{2}$ yrs	10 yrs	15 yrs	APR
£1000	34.96	24.11	18.92	16.50	N/A	16.7%
£3000	104.88	72.32	56.75	49.51	N/A	16.7%
£6000	N/A	140.86	109.39	94.61	81.52	15.3%
£10000	N/A	231.03	178.27	153.35	131.03	14.5%
£15000	N/A	346.54	267.41	230.02	196.55	14.5%

Special Scheme for Recent Purchasers---- *Sorry No Tenants*

6 Find the monthly repayments on a loan of £3000 over 5 years.

7 Find the total payable on a loan of £6000 over 7$\frac{1}{2}$ years.

8 What does it cost in interest in settling a loan of £10 000 over 10 years?

9 How much would the repayments be for a loan of £9000 over 7$\frac{1}{2}$ years?

10 What could be the total repayments on a loan of £4000 over 5 years?

Investigation E

Inside many national and local newspapers you will find advertisements for loans. Cut these out, and compare the interest rates and costs of taking out loans. Can you reach any conclusions about them? Are there any conditions attached to any of these loans? Why do you think such conditions are needed?

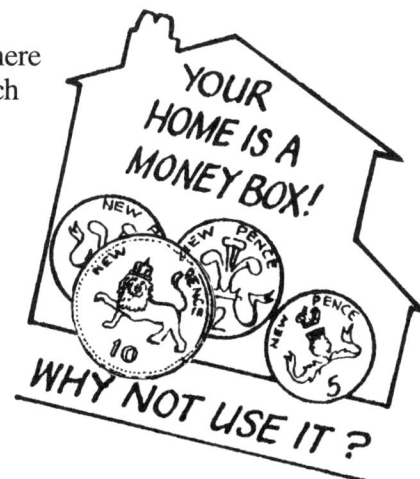

YOUR HOME IS A MONEY BOX!

WHY NOT USE IT ?

3. Algebra (1)

Introduction

Algebra uses letters or other symbols in place of numbers, in order to give rules for working out results. From these rules, numerical answers can be found.

In this triangle, the perimeter (distance round the outside) is

$$4+4+4=12 \text{ cm}$$
$$\text{or} \quad 3\times4=12 \text{ cm}$$

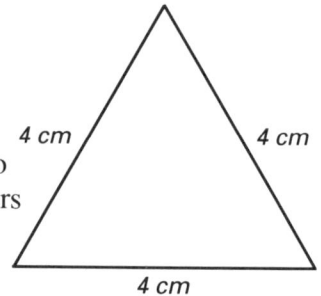

In any equilateral triangle, if we call the length of the side s cm, then the perimeter will be

$$s+s+s \text{ or } 3\times s \text{ cm}$$

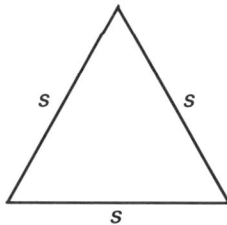

In algebra, we usually miss out the $\times$ sign. So instead of $3\times s$, we write $3s$. This still means '3 times s' or '3 lots of s'. For example, if the length of the side of the triangle is 10 cm, then $s=10$, and the perimeter is $3\times10=30$ cm.

Let us begin by putting numbers instead of letters in simple algebraic expressions.

Example 1

Work out the value of $4k$ when $k=7$.

$4k$ means $4\times k$. If $k=7$, then

$$4\times k=4\times7$$
$$=28$$

Example 2

If $m=3$, what is the value of $8m+1$?

$8m+1$ means work out the value of $8\times m$, then add 1.

$$8\times m=8\times3=24$$
$$\text{So} \quad 8m+1=24+1$$
$$=25$$

EXERCISE **3.1**

Work out the value of each of these expressions.

1 $5a$, when $a = 4$ **2** $3t$, when $t = 8$

3 $7c$, when $c = 50$ **4** $2k$, when $k = 30$

5 $6l$, when $l = 5$ **6** $3x + 2$, when $x = 7$

7 $8s + 3$, when $s = 2$ **8** $4p + 5$, when $p = 20$

9 $6d + 18$, when $d = 3$ **10** $100f + 400$, when $f = 6$

More than one letter

The perimeter of this rectangle is

$$7 + 7 + 4 + 4 = 14 + 8$$
$$= 22 \text{ cm}$$

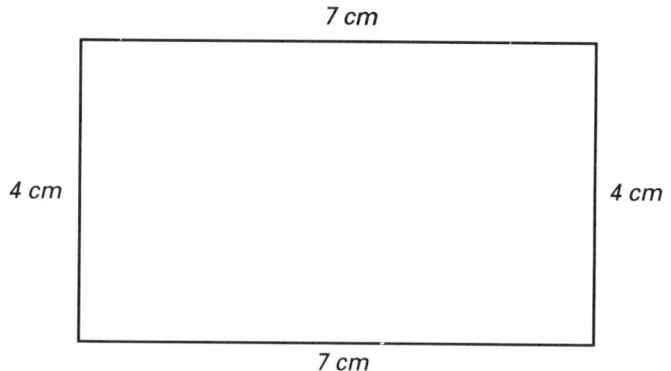

In this rectangle, the perimeter will be

$$b + b + 3 + 3$$

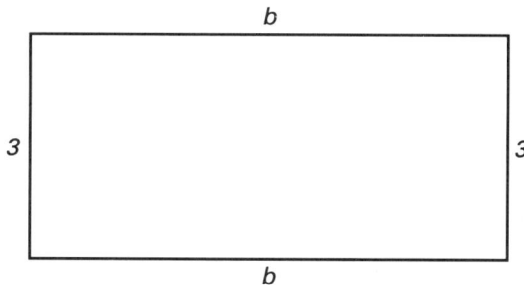

We cannot make this into a single 'answer', like 22 cm, but we can write the answer more concisely as $b + b + 6$ or $2b + 6$ (remember, $2b = 2 \times b$).

Example 3

Write an expression for the perimeter of this rectangle.

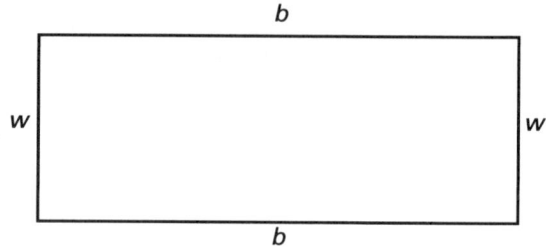

Perimeter $= b + b + w + w$
$\qquad = 2b + 2w$

Example 4

What is the perimeter of this pentagon?

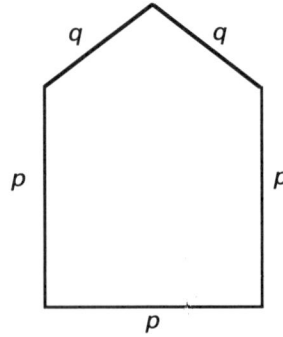

Perimeter $= p + p + p + q + q$
$\qquad = 3p + 2q$

In each of these examples, we cannot go further until we know the actual values of the letters.

For example, if $p = 8$ cm and $q = 5$ cm in Example 4 above, then the perimeter is

$$(3 \times 8) + (2 \times 5) = 24 + 10$$
$$= 34 \text{ cm}$$

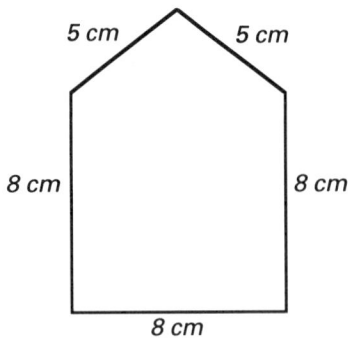

Example 5

Write $x + x + 3y + x + 2y$ as simply as you can.

Taking the x's first, we can write this as $x + x + x + 3y + 2y$, which simplifies to $3x + 5y$.

EXERCISE 3.2

Write these expressions as simply as you can.

1 $x+x+7$ **2** $2a+6+a+8$

3 $p+p+q+2p+3q$ **4** $3t+6t+5$

5 $2b+7+b$ **6** $8+5k+6+4k$

7 $4c+5+6c+7$ **8** $z+2+z+4+5z$

9 $3+2n+n+7+4n$ **10** $2s+8+s+s+1$

11 $p+p+q$ **12** $3x+2y+4x$

13 $m+3n+4m+5n$ **14** $6a+2a+3b+4b+1$

15 $8c+5d+2c+d$ **16** $y+z+9z+2y$

17 $3t+6u+2t+7$ **18** $150+7h+50+7h$

19 $8p+2q+11p+3$ **20** $1+2x+3+4x+5+6x$

Subtraction

The same general rule for addition applies to subtraction.

Example 6

Simplify these expressions:

(a) $4x-x$ (b) $7y-4y$ (c) $a+6-a$ (d) $2x+y-5x$

(a) $4x-x=3x$ (b) $7y-4y=3y$
(c) $a+6-a=6$ (d) $2x+y-5x=y-3x$

EXERCISE 3.3

Simplify these expressions.

1 $7t-2t$ **2** $3a-a$

3 $3x-2x+5$ **4** $7k-3k-3$

5 $n+5-n-4$ **6** $4y+2-y-3$

7 $8p+3p-5-5p$ **8** $3v+7-2v+2-v$

9 $4-x-1-3x$

10 $20-4c-5c+6c$

11 $4b+3c-2b-2c$

12 $17x-12y-4x+6y$

13 $10t+3u-3t+10u$

14 $p-2q-3q+2p$

15 $7a-2b-6a+3b$

16 $5x+y-3-y-3x$

17 $6m+8-3m-7-2m$

18 $14-3x-7x$

19 $5-4h-3+9h-2$

20 $a+2b-3c-a-3b+2c$

EXERCISE 3.4

Here are some expressions for you to work out, in the same way as you did in Exercise 3.1. The letters could stand for any type of unit (kilograms, metres, pence, etc.); just work out the value as a number.

1 $4x+6$, when $x=7$

2 $3a+7$, when $a=10$

3 $t-4$, when $t=21$

4 $6x+y$, when $x=3$ and $y=2$

5 $12-k$, when $k=5$

6 $3g+14h$, when $g=8$ and $h=3$

7 $5c-3d$, when $c=11$ and $d=3$

8 $100-7b$, when $b=9$

9 $3t-15$, when $t=5$

10 $6y+8z$, when $y=6$ and $z=8$

EXERCISE 3.5

For this exercise you will need your calculator. The letters stand for £, so give your answers in £ and pence.

1 $4k+£15$, when $k=£2.43$

2 $12d-£15.50$, when $d=£8.72$

3 $5y-£105.30$, when $y=£38.98$

4 $16z+£14.94$, when $z=£4.48$

5 $£300-3t$, when $t=£69.99$

6 $£676.48+r$, when $r=£492.04$

7 $£10.50+5s$, when $s=£12.75$

8 $6a-£0.84$, when $a=£0.91$

9 $£20+7m$, when $m=£25$

10 $£967.21-2w$, when $w=£87.49$

Multiplication

Look at this rectangle.

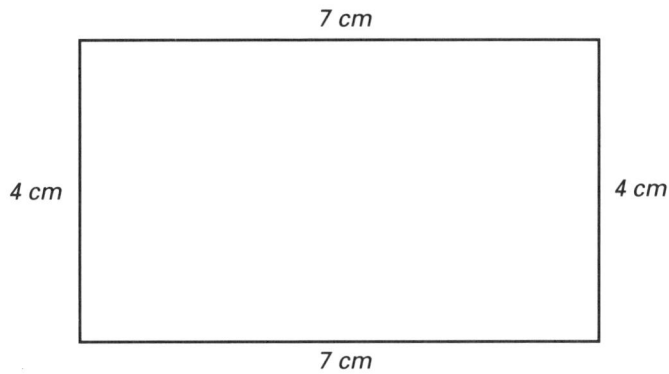

Its area is $7 \times 4 = 28$ cm². (Remember that cm² is shorthand notation for 'square centimetres'.)

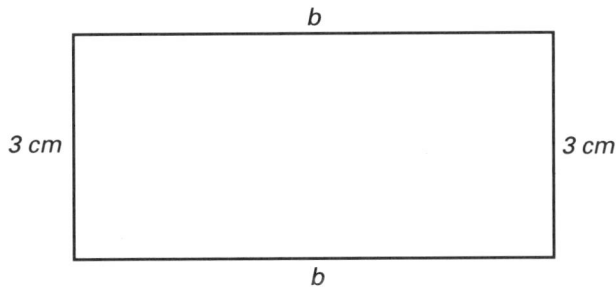

Similarly, the area of this rectangle is $b \times 3$ or $3b$ cm². (We usually write numbers first followed by letters, so it is usual to write $3b$ rather than $b3$.)

If the breadth $b = 5$ cm, then

$$\begin{aligned} \text{area} &= 3 \times 5 \\ &= 15 \text{ cm}^2 \end{aligned}$$

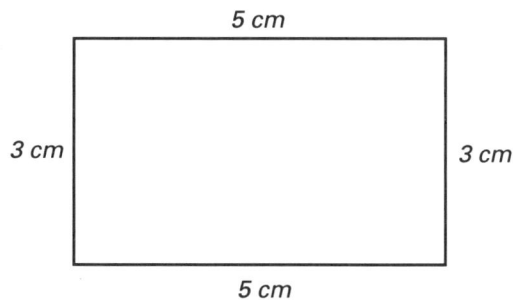

This rectangle has an area of $b \times w$, or bw cm².

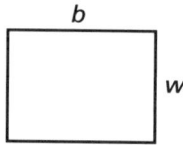

Four of these rectangles have an area of $4 \times b \times w$ or $4bw$ cm².

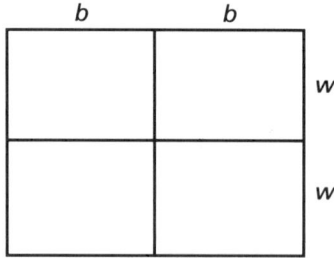

If the breadth b is 8 cm and the width w is 5 cm, then the area is $4 \times 8 \times 5 = 160$ cm².

Example 7

What is the area of this rectangle?

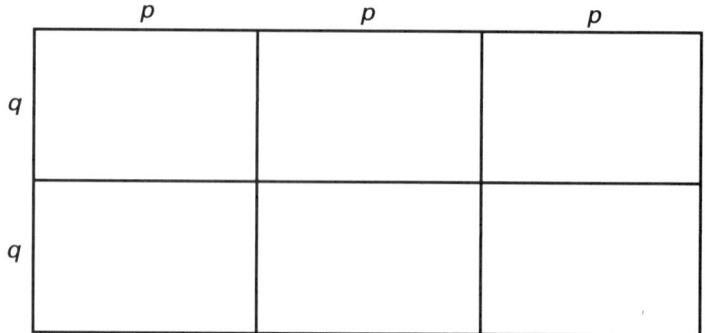

There are six rectangles, each with an area of $p \times q$ or pq cm², so the total area is $6pq$ cm².

We could equally well say that the area is $(3p \times 2q)$ cm², because the length is $3p$ cm and the width is $2q$ cm.

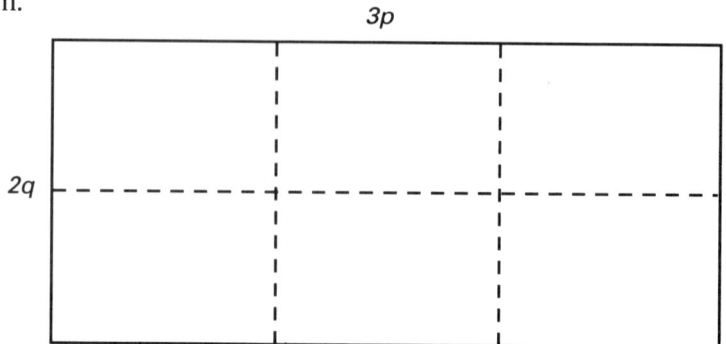

We can write this as $3 \times p \times 2 \times q$; if we write the numbers first, this is

$$3 \times 2 \times p \times q = 6 \times p \times q$$
$$= 6pq \text{ cm}^2 \quad \text{as before.}$$

Example 8

Write these terms more simply:
(a) $7p \times q$ (b) $6m \times 9n$.

(a) $7p \times q = 7pq$

(b) $6m \times 9n = 6 \times m \times 9 \times n$
$$= 6 \times 9 \times m \times n$$
$$= 54mn$$

EXERCISE 3.6

Simplify these terms.

1 $3a \times 5b$

2 $6f \times g$

3 $10i \times 10j$

4 $8 \times 2y \times 2z$

5 $4p \times 6q$

6 $3 \times 3u$

7 $2m \times 3n \times 4$

8 $20p \times 5q$

9 $a \times b \times c$

10 $2 \times a \times t$

Example 9

What is the value of $5a \times 2b$, when $a = 8$ and $b = 10$?

$$5a \times 2b = 5 \times 2 \times a \times b$$
$$= 10 \times a \times b \quad \text{(or } 10ab\text{)}$$
$$= 10 \times 8 \times 10$$
$$= 10 \times 10 \times 8$$
$$= 800$$

Example 10

Work out the value of $3pq - 4p$, when $p = 5$ and $q = 2$.

Each term needs to be worked out separately, then the subtraction done.

$$3pq = 3 \times 5 \times 2 = 30 \qquad 4p = 4 \times 5 = 20$$

So $3pq - 4p = 30 - 20$
$$= 10$$

EXERCISE 3.7

Work out the value of each of these terms (you will need a calculator for questions 11–20).

1 $10at$, when $a = 10$ and $t = 4$.

2 $3pq$, when $p = 5$ and $q = 7$.

3 $6uv$, when $u = 1$ and $v = 2$.

4 $5rs + 6s$, when $r = 4$ and $s = 2$.

5 $2pr \times h$, when $p = 3$, $r = 10$ and $h = 4$.

6 $u + at$, when $u = 20$, $a = 10$ and $t = 4$.

7 $7ax - 2ay$, when $a = 3$, $x = 3$ and $y = 10$.

8 $100 - 8p - 3pq$, when $p = 6$ and $q = 2$.

9 $ab + bc + ca$, when $a = 2$, $b = 4$ and $c = 6$.

10 $36 + 3ij - 12j$, when $i = 2$ and $j = 6$.

11 xy, when $x = 12.3$ and $y = 15$.

12 $4rs$, when $r = 3.5$ and $s = 1.4$.

13 $6ab + 8b$, when $a = 7.3$ and $b = 9.7$.

14 $500 - 6pq$, when $p = 17.6$ and $q = 2.7$.

15 $kC + 32$, when $k = 1.8$ and $C = 20$.

16 $9w - 2k$, when $w = 34.7$ and $k = 68.3$.

17 $3xy + 5y - 7$, when $x = 2.53$ and $y = 4.5$.

18 $76.9 + c - 2ct$, when $c = 35.9$ and $t = 0.48$.

19 $100 - 5abc$, when $a = 1.25$, $b = 4$ and $c = 4$.

20 $xy + yz + zx$, when $x = 6.4$, $y = 16.2$ and $z = 5.3$.

Division

Example 11

If the area of the carpet is 24 m², and its width is 3 m, how long is it?

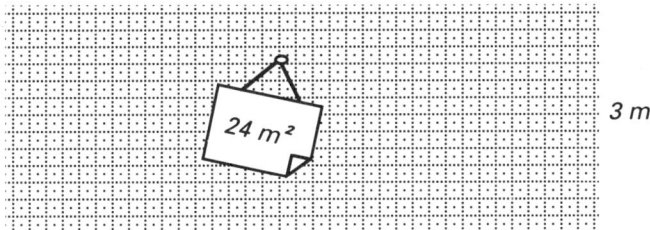

We need to work out

$$24 \div 3 = 8$$

The carpet is 8 m long.

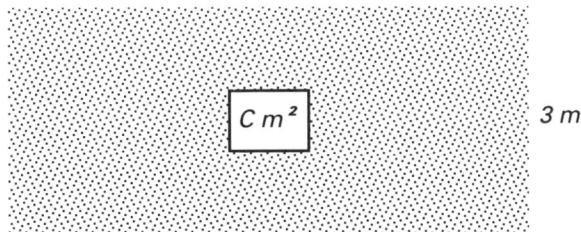

If the area of the carpet had been C square metres, then the length would have been $\frac{1}{3}$ of C or $(C \div 3)$ m. (We usually write $\frac{C}{3}$ m. In Example 10, we could have written $\frac{24}{3} = 8$ m.)

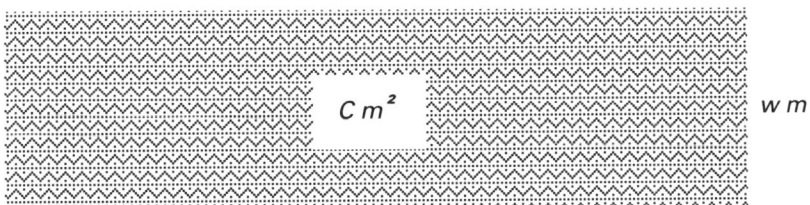

Also, if the width had been w m, then the length would have been $\frac{C}{w}$.

EXERCISE 3.8

Work out the value of these terms.

1 $\dfrac{c}{3}$, when $c = 12$

2 $\dfrac{20}{k}$, when $k = 4$

3 $5 + \dfrac{t}{7}$, when $t = 21$

4 $\dfrac{M}{5} - 6$, when $M = 50$

5 $\dfrac{45}{r}$, when $r = 9$

6 $40 - \dfrac{v}{9}$, when $y = 90$

7 $4p - \dfrac{q}{2}$, when $p = 4$ and $q = 6$

8 $16 + 5w - \dfrac{w}{4}$, when $w = 12$

9 $2h + \dfrac{h}{3} + 4$, when $h = 24$

10 $\dfrac{x}{2} + \dfrac{x}{3} + \dfrac{x}{4}$, when $x = 36$

Investigation A

(a) On your calculator, work out $\dfrac{100}{5 \times 4}$

The answer is 5, as $5 \times 4 = 20$, and $\dfrac{100}{20} = 5$.

If you pressed $100 \div 5 \times 4$, your calculator would give 80, because it would have worked out $100 \div 5$, which is 20, and then *multiplied* by 4 to give 80, which is the wrong answer.

(b) Investigate which buttons you would press, in which order, to work out the value of $\dfrac{p}{q \times r}$. Use the memory store if you wish, and check your method by choosing suitable numbers for p, q and r.

Now try to find any other sequences of operations which will give the correct answer, each time checking by putting in suitable numbers.

(c) Repeat (b), using the term $\dfrac{p \times q}{r \times s}$

Example 12

Work out $\dfrac{3p}{4}$, when $p = 8$.

We can consider this expression in two ways:

(a) as a quarter of $3p$

$= \dfrac{1}{4}$ of (3×8)

$= \dfrac{1}{4}$ of 24

$= 6$

or (b) 3 times a quarter of p

$= 3 \times (\dfrac{1}{4}$ of 8)

$= 3 \times 2$

$= 6$

Example 13

What is $\dfrac{4a}{7b}$, when $a=35$ and $b=2$?

We can think of the expression as:

(i) $\dfrac{4\times a}{7\times b}$ (ii) $\dfrac{1}{7}$ of $\dfrac{4\times a}{b}$ (iii) $\dfrac{1}{7\times b}\times 4\times a$.

All these are *equivalent*, and will therefore give the same value when worked out. Some ways of working, however, are much simpler than others.

(i) $\dfrac{4\times a}{7\times b}=\dfrac{4\times 35}{7\times 2}=\dfrac{140}{14}=10$

(ii) $\dfrac{1}{7}$ of $\dfrac{4\times a}{b}=\dfrac{1}{7}$ of $\dfrac{4\times 35}{2}=\dfrac{1}{7}\times\dfrac{140}{2}=\dfrac{1}{7}\times 70=10$

(iii) $\dfrac{1}{7\times b}\times 4\times a=\dfrac{1}{14}\times 4\times 35=\dfrac{1}{14}\times 140=10$

Example 14

Work out $\dfrac{60}{rs}$, when $r=2$ and $s=6$.

The denominator is $r\times s=2\times 6=12$.

So $\dfrac{60}{rs}=\dfrac{60}{12}=5$

EXERCISE 3.9

When trying these questions, the method you choose is likely to depend upon the numerical values of the letters.

1 $\dfrac{2p}{3}$, when $p=10$.

2 $\dfrac{3k}{10}$, when $k=50$.

3 $\dfrac{5a}{6}+\dfrac{2b}{9}$, when $a=12$ and $b=18$.

4 $\dfrac{20}{x}-\dfrac{12}{x}$, when $x=2$.

5 $3d-\dfrac{20}{d}+32$, when $d=10$.

6 $\dfrac{48}{ab}+\dfrac{60}{a}-\dfrac{28}{b}$, when $a=6$ and $b=4$.

7 $\dfrac{5p}{r} + \dfrac{54}{qr} + \dfrac{4q}{p}$, when $p = 12$, $q = 9$ and $r = 6$.

8 $abc - \dfrac{50}{5b}$, when $a = 4$, $b = 5$ and $c = 6$.

9 $4 + \dfrac{x}{4} + \dfrac{4}{x}$, when $x = 4$.

10 $\dfrac{3p}{q} + \dfrac{10p}{r}$, when $p = 16$, $q = 6$ and $r = 80$.

Indices

Example 15

What is the area of this square?

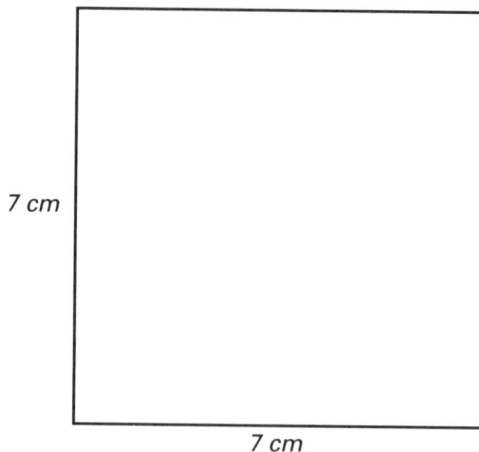

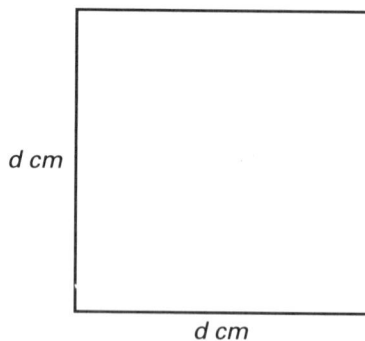

The area of this square is $7 \times 7 = 49$ cm².

7 cm

7 cm

Example 16

What is the area of this square?

The area of this square is $d \times d$ or dd cm².

d cm

d cm

When we multiply a number by *itself* we say we are **squaring** the number.
(Why do you think we call it 'squaring'?).
So *dd* is '*d* squared', and we write it in a special way: d^2.

Example 17

Work out (a) t^2, when $t = 8$ (b) $4c^2$, when $c = 5$
(c) $7a^2$, when $a = 3$ (d) $p^2 - q^2$, when $p = 10$ and $q = 4$.

(a) $t^2 = t \times t = 8 \times 8 = 64$.
(b) $4c^2 = 4 \times c \times c = 4 \times 5 \times 5 = 4 \times 25 = 100$.
You must take care with terms like $4c^2$. It is *not* $4 \times c$, then squared. It is 4 lots of c^2; the c must be squared first, and then that answer multiplied by 4.
(c) $7a^2 = 7 \times a^2 = 7 \times 3 \times 3 = 7 \times 9 = 63$.
(d) $p^2 - q^2 = 10 \times 10 - 4 \times 4 = 100 - 16 = 84$.

EXERCISE 3.10

Work out the value of each of these expressions.

1 p^2, when $p = 3$
2 $2q^2$, when $q = 5$
3 $k^2 - 12$, when $k = 6$
4 $3x^2 + 2x$, when $x = 4$
5 $m^2 - 6m + 9$, when $m = 7$
6 $10t^2 - 2u^2$, when $t = 2$ and $u = 3$
7 $2s^2 + 3t$, when $s = 6$ and $t = 2$
8 $200 - 3y^2$, when $y = 8$
9 $n^2 + 81$, when $n = 9$
10 $a^2 + b^2 - ab$, when $a = b = 10$

Example 18

What is the volume of this cube?

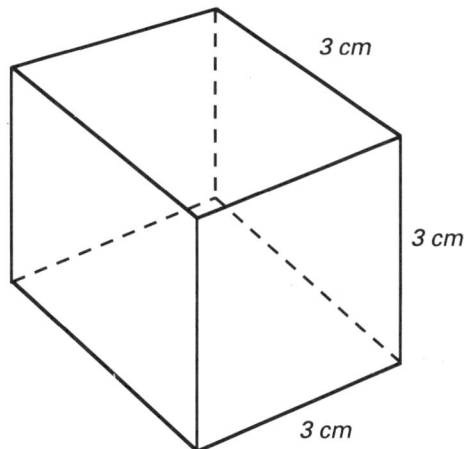

3 cm
3 cm
3 cm

Volume $= 3 \times 3 \times 3 = 27 \text{ cm}^3$.
(Remember that cm^3 is shorthand notation for 'cubic centimetres'.)

Example 19

What is the volume of this cube?

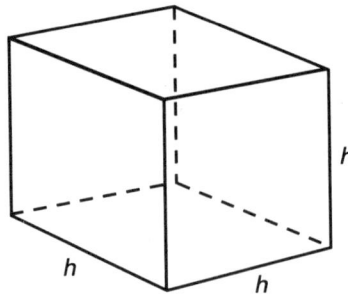

Volume $= h \times h \times h$

Following the same principle as for 'squaring', we can write:

$$h \times h \times h = h^3$$

h^3 is called 'h cubed'. (This is the same as $h^2 \times h$ or $h \times h^2$.)

Extending this idea: $p \times p \times p \times p = p^4$ and $b \times b \times b \times b \times b = b^5$. We say p^4 is 'p to the power of 4', and b^5 is 'b to the power of 5'. The little number (the 4 or the 5 in these two examples), written just above and to the right, is called the power or **index** of the letter. The plural of index is 'indices' (the heading of this section of work).

If $p = 3$, then $p^4 = 3 \times 3 \times 3 \times 3 = 81$, and if $b = 2$, then $b^5 = 2 \times 2 \times 2 \times 2 \times 2 = 32$.

Example 20

Work out $3k^4$, when $k = 2$.

As before, $3k^4 = 3$ lots of k^4. This means that you need to work out k^4, *then* multiply the answer by 3. So

$$3k^4 = 3 \times (2 \times 2 \times 2 \times 2) = 3 \times 16 = 48$$

Example 21

Simplify (a) $v^3 \times v^2$ (b) $5t^3 \times 3t$.

(a) $v^3 \times v^2 = v \times v \times v \ \times \ v \times v = v^5$.

(b) $5t^3 \times 3t = 5 \times t \times t \times t \times 3 \times t = 5 \times 3 \times t^4 = 15t^4$.

EXERCISE 3.11

Simplify each expression.

1 $a^2 \times a$

2 $2t^3 \times t^2$

3 $5y \times 4y^2$

4 $6p^2 \times 2p^3$

5 $p^2 \times q^3 \times p$

6 $5s^3 \times 6t^4$

7 $a^2 \times b^5 \times ab$

8 $2d^2 \times 3d^3 \times 4d^4$

9 $5 \times 8y \times z^3$

10 $9h^5 \times 2h^2$

Investigation B

We can write a^3 as $a \times a \times a$ or $a \times a^2$ or $a^2 \times a$, i.e. in three different ways.

Investigate the number of different ways of writing a^4 (there are 7). Now repeat with a^5, a^6 and a^7.

Set your results out in a table, starting with a^2, then a^3, then a^4, etc.

Example 22

Simplify (a) $\dfrac{n^2}{n}$ (b) $\dfrac{3u^2v^2}{v^3}$ (c) $\dfrac{6p^4q^2}{2pq^2}$.

(a) $\dfrac{n^2}{n} = \dfrac{n \times n}{n} = n$

(b) $\dfrac{3u^2v^2}{v^3} = \dfrac{3 \times u \times u \times v \times v}{v \times v \times v} = \dfrac{3u^2}{v}$

(c) $\dfrac{6p^4q^2}{2pq^2} = \dfrac{6 \times p \times p \times p \times p \times q \times q}{2 \times p \times q \times q} = 3p^3$

As in multiplication, we write the numbers first, then the letters.

EXERCISE 3.12

Simplify each of these expressions.

1 $\dfrac{x^4}{x}$

2 $\dfrac{4k^2}{k}$

3 $\dfrac{m^2}{3m}$

4 $\dfrac{12t^3}{3t^2}$

5 $\dfrac{7qr^3}{14qr^3}$

6 $\dfrac{x^2y^2z^2}{xy^2z^3}$

7 $\dfrac{2t^5}{6u^2}$

8 $\dfrac{30s}{6s^3}$

9 $\dfrac{8ab^3}{2a^3b}$

10 $\dfrac{36xy^2}{6xy}$

A word of warning

$3a^2 \times 4a^3 = 12a^5$, but you *cannot* simplify something like $3a^2 + 4a^3$ by adding, as it is like trying to add an area to a volume. Only when we know the actual value of a can we do any simplifying.

Example 23

Simplify $x^2 - 4x + 3x^2 + x$.

Taking the x^2 terms first, then the x terms:

$$x^2 + 3x^2 - 4x + x = 4x^2 - 3x.$$

EXERCISE 3.13

Simplify these expressions.

1 $x^2 + x + 3x$

2 $3k^3 + 5k - k$

3 $2t^2 + 3t^3 + 4t^2 - 5t^3$

4 $m^5 + 3n + 3m^5 - n$

5 $7n^2 - \dfrac{4n^3}{n}$

6 $c + 4c^2 - 2c$

7 $4r^2 - 3r - r^2 + 2r$

8 $\dfrac{6a^4}{2a} + 4a^3$

9 $3x^2 + 4xy - 5xy + 3x^2$

10 $3v^3 + 4v^2 - 3v^3 - 3v^2$

Brackets

This rectangle is made up of two smaller ones.

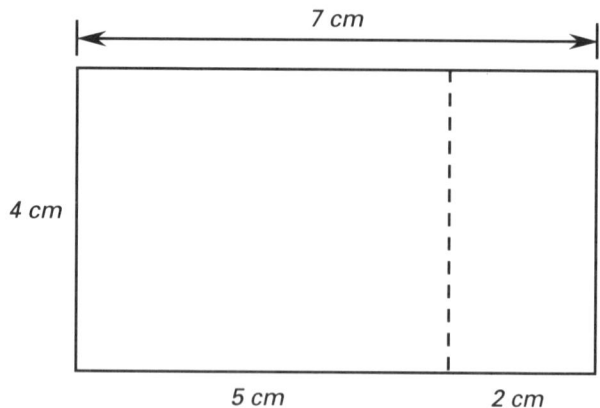

Its area is

$$4 \times (5 + 2) = 4 \times 7 = 28 \text{ cm}^2.$$

The brackets are there to tell you that you have to work out $5 + 2$ first, then multiply by 4.

We could have worked out the area of each of the smaller rectangles, and then added these areas.

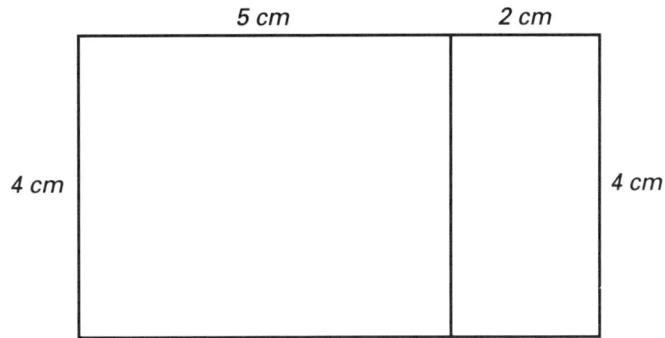

$(4 \times 5) + (4 \times 2) = 20 + 8 = 28$ cm².

Again, the brackets show which parts have to be worked out first.

When we know all the dimensions, then the first method is the one we would invariably use. If we don't know all the dimensions, however, we have to use the second approach, splitting the area into two parts.

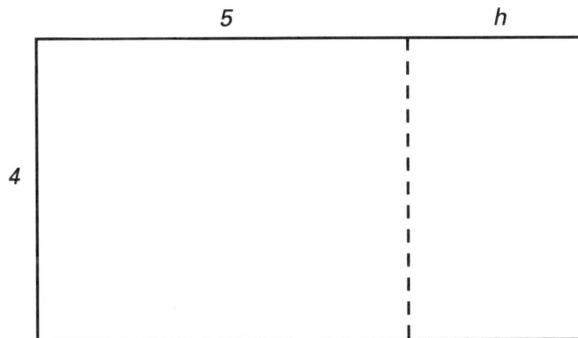

Here, we can say that the area is

$$4 \times (5 + h) \quad \text{or} \quad \underline{4(5 + h)}$$

We can also write the area as

$$A + B = (4 \times 5) + (4 \times h)$$
$$= \underline{20 + 4h}$$

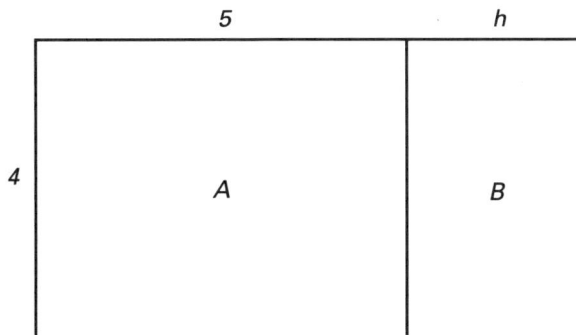

As these underlined terms are both equal to the area of the rectangle, then we can say that

$$4(5+h)=20+4h \quad (4 \text{ lots of } 5+4 \text{ lots of } h).$$

In general, we can remove the brackets from an expression by multiplying each term inside the bracket by the term immediately in front of the bracket.

Example 24

Express the following without brackets:
(a) $3(z+8)$ (b) $7(t-3)$ (c) $4(3a+2b)$ (d) $3(d^2-4)$.

(a) $3(z+8)=(3 \times z)+(3 \times 8)=3z+24$

(b) $7(t-3)=(7 \times t)-(7 \times 3)=7t-21$

(c) $4(3a+2b)=(4 \times 3a)+(4 \times 2b)=12a+8b$

(d) $3(d^2-4)=(3 \times d^2)-(3 \times 4)=3d^2-12$

EXERCISE 3.14

Write these expressions without brackets.

1 $2(x+3)$

2 $5(r-6)$

3 $3(f-3)$

4 $8(2s+3)$

5 $6(3y+4)$

6 $5(5-z)$

7 $12(x^2+y^2)$

8 $3(5a+2b)$

9 $2(7h-5h^3)$

10 $10(10d^2-1)$

When there is a letter outside a bracket the same rule applies.

Example 25

Rewrite the following without brackets:
(a) $t(t^2+4)$ (b) $m(2m-3n)$ (c) $3d(5+d)$ (d) $2a^3(3a-4)$.

(a) $t(t^2+4)=t^3+4t$

(b) $m(2m-3n)=2m^2-3mn$

(c) $3d(5+d)=15d+3d^2$

(d) $2a^3(3a-4)=6a^4-8a^3$

EXERCISE 3.15

Write these expressions without brackets.

1 $k(k-3)$ **2** $c(3c+4)$

3 $b(4-9b^2)$ **4** $s(6s+7t)$

5 $6r(r+h)$ **6** $2t(t^3-3t)$

7 $x(3y-4z)$ **8** $4z^2(5-6z)$

9 $5pq(p+2q)$ **10** $3a^2(b^2+c^2)$

Factorising

The opposite process to expanding by multiplying in order to remove brackets is **factorising**. Factorising is trying to find some number or letter which is common to two (or more) terms; in other words, a *common factor* of those terms. The process of factorising will involve us in having to put in brackets, instead of getting rid of them.

Example 26

Factorise $3z+24$.

Look back at Example 24(*a*). We can see that $3z+24$ is the same as $3(z+8)$. What we need to look for is some number (or letter) which is common to both terms. In this case it is 3, because each term has a *factor* of 3.

$$3z \quad + \quad 24$$
$$\downarrow \qquad\qquad \downarrow$$
$$3\times z+3\times 8$$

$3z+24 = (3\text{ lots of } z)+(3\text{ lots of } 8) = 3\text{ lots of } (z+8) = 3(z+8)$.

Example 27

Factorise (*a*) $8t-4$ (*b*) $3a^2+9a$ (*c*) pq^2+p^2q.

(*a*) $8t-4 = 4(2t-1)$, as 4 is the only factor of *both* terms.

(*b*)   Both 3 and *a* will divide exactly into each term, so we can take out the common factor of 3*a*,

     $3a^2+9a = 3a(a+3)$

(*c*)   Again, as both *p* and *q* are factors of each term, we can write

     $pq^2+p^2q = pq(q+p)$

EXERCISE 3.16

Factorise these expressions, taking care to find all the factors common to each term.

1 $10s - 5$ **2** $4r + 12$

3 $p^3 + 2p$ **4** $3a^2 - 15ab$

5 $6x + 9xy$ **6** $8m - 10n$

7 $6pq - 3q^2$ **8** $16 + 24g$

9 $100n + 10n^2$ **10** $12d^3 + 21d^2$

4. Algebra (2)

Substituting numbers and letters for words

In Book 3X, Chapter 4, we looked at formulas. As a reminder, a formula is merely a rule for working out some value, using numbers and letters in place of words.

As an example, if I earn £2 each night on my evening paper round, how much will I earn in six days, or in fifteen days?

We could write: $E = £2 \times n$, where E stands for earnings, and n stands for the number of evenings that I do my paper round. So in a week in which I do six rounds, n will be 6, and

$$E = £2 \times 6 = £12$$

In fifteen days, my earnings will be £2 × 15 = £30.

The rule or **formula** is $E = £2 \times n$

Example 1

The instructions for working out the time required to cook a joint of meat say 'allow 40 minutes per kilogram, plus 30 minutes'. Write this information as a formula.

We need to choose some letters to enable us to write down a formula.

First, give a single letter to the thing that you are asked to find – in this case the time taken to cook the meat. Let us choose T to stand for the cooking time in minutes. From the instructions, the time depends only on the weight of the joint. If we let w be the weight of the joint, in kilograms, then we can say that

cooking time $(T) = (40 \times w)$ minutes plus 30 minutes
$$T = 40w + 30$$

So a 1.5 kg joint will take

$$(40 \times 1.5) + 30 = 60 + 30$$
$$= 90 \text{ minutes} \quad \text{or} \quad 1\tfrac{1}{2} \text{ hours}$$

49

Example 2

To change degrees Celsius into degrees Fahrenheit, multiply the Celsius temperature by 1.8 and then add 32.

(*a*) Write this as a formula.
(*b*) Work out the equivalent of 10°C and 30°C in degrees Fahrenheit.

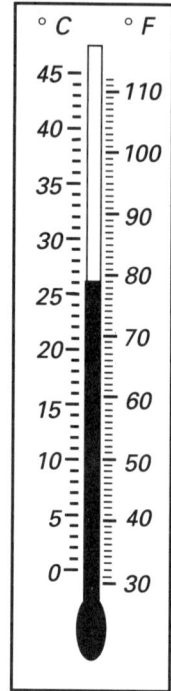

We need to write a formula for the Fahrenheit temperature.
(*a*) If we let $F=$ Fahrenheit temperature, and $C=$ Celsius temperature, then the formula is $F=1.8 \times C + 32$, or $F=1.8C+32$.
(*b*) If $C = 10°C$, then $F = (1.8 \times 10) + 32$
$$= 18 + 32$$
$$= 50°F$$

If $C = 30°C$, then $F = (1.8 \times 30) + 32$
$$= 54 + 32$$
$$= 86°F$$

EXERCISE 4.1

1 When I babysit, I charge 50p, plus 75p an hour.

(*a*) Write a formula for my earnings (£E) if I babysit for h hours.

(*b*) How much will I earn if I babysit for four hours?

2 The cost of gravel is £6.50 per tonne, plus £8.00 delivery charge.

(*a*) Write a formula for the cost (£C) of having k tonnes delivered.

(*b*) I need 20 tonnes of gravel for my drive. How much will this cost me?

3 My cricket club charges £2 for each game in which I play, plus a joining fee of £12.

(*a*) Write a formula for the total fees due (£F) if I play g games.

(*b*) In my first season I play 18 games. How much will I have to pay the cricket club altogether?

4 An examination board charges 60p per copy for old examination papers, plus a postage and packing charge of £1.05.

(*a*) Write a formula for the cost (£*X*) of buying *n* examination papers. (**Take care**, 60p needs to be changed to £.)

(*b*) The fourth year needs 80 copies of last year's examination paper. How much will it cost to buy them?

5 My mum said 'For each sum you get right, I'll give you 8p. For each sum you get wrong, you must give me 5p'.

(*a*) Write a formula for my profit (*P*) if I get *r* sums right and *w* sums wrong.

(*b*) On a day when I got ten sums right, I made no profit (or loss). How many sums did I get wrong?

6 My telephone bill each quarter is made up of a standard charge of £16.55, plus £4.40 for every 100 units used.

(*a*) Write a formula for my telephone bill (£*B*) for a quarter during which I use *x* hundred units.

(*b*) How much will my bill be for a quarter during which I use 1500 units?

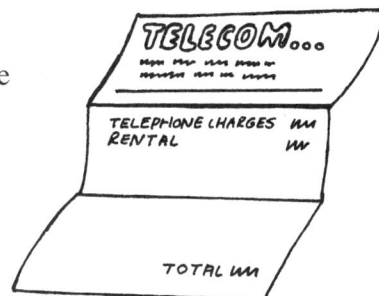

7 For painting a house, a man charges £10.00 per hour, *less* the cost of the paint.

(*a*) Find the cost (£*C*) of painting a house which takes *h* hours, for which the paint costs £*D*.

(*b*) How much will I have to pay the man if it takes him 12 hours to paint the house and the paint costs me £18?

8 There are 50 foreign stamps in a packet. As long as I buy at least six packets, the shopkeeper will give me an extra packet containing 75 stamps.

(*a*) Write a formula for the number of stamps (*N*) that I will have if I buy *x* packets of stamps (assuming that I buy at least six packets).

(*b*) I can afford to buy nine packets of stamps. How many stamps will I receive from the shopkeeper?

9 On holiday, I can hire a bicycle for a deposit (non-returnable) of £25, plus £5 per day.

　(*a*) Write a formula for the cost (£*K*) of hiring a bicycle for *d* days.

　(*b*) If I hire the bicycle for nine days, how much will I have to pay?

10 My uncle gave me £20 on condition that I spent it on wool for my knitting machine.

　(*a*) If a ball of wool costs £1.50, write a formula for the amount of money (£*S*) that I have left after buying *b* balls of wool.

　(*b*) How much will I have left if I buy nine balls of wool?

　(*c*) What is the greatest number of balls of wool I could buy with the money which my uncle gave me?

Evaluating formulas

In the next exercise you are given a formula, or rule, and you have to work out a value by substituting numbers for letters.

In order to reduce errors, and to help you clearly through the problem, it is necessary to tackle separately the two distinct stages of (i) **substitution** and (ii) **evaluation**.

Doing both stages at the same time can easily lead to errors, even if the formula is quite simple.

Example 3

In the formula $t = \dfrac{v-u}{a}$, what is the value of *t* when $u = 5$, $v = 23$ and $a = 3$?

(i) Substituting numbers for letters: $t = \dfrac{23-5}{3}$.

(ii) Evaluating: $t = \dfrac{18}{3} = 6$.

Example 4

The volume of a square based pyramid is given by the formula $V = \dfrac{x^2 h}{3}$,

where x is the length of the side of the square base, and h is the perpendicular height of the pyramid. What is the volume of a pyramid in which $x = 6$ m and $h = 8$ m?

(i) Substituting: $V = \dfrac{6 \times 6 \times 8}{3}$

(ii) Evaluating: $V = \dfrac{36 \times 8}{3} = 12 \times 8 = 96 \text{ m}^3$.

(We could have worked out $36 \times 8 = 288$, then divided by 3 to give 96.)

Example 5

The total surface area of a cylinder of height h and radius r is given by the formula $A = 2\pi r(r + h)$.

Calculate the total surface area of a cylinder of height 15 cm and radius 5 cm, given that $\pi = 3.14$.

(i) Substituting: $A = 2 \times 3.14 \times 5(5 + 15)$
(ii) Evaluating: taking the bracket first, $5 + 15 = 20$, so

$$\begin{aligned} A &= 2 \times 3.14 \times 5 \times 20 \\ &= 2 \times 3.14 \times 100 \\ &= 6.28 \times 100 \\ &= 628 \text{ cm}^2 \end{aligned}$$

EXERCISE 4.2

Work out the left-hand letter in each formula by replacing the right-hand letters by the numbers given and evaluating. You may need your calculator for some of the questions.

1 $p = 2(a + b)$, when $a = 7$ and $b = 5$.

2 $y = mx + c$, when $m = 2$, $x = 12$ and $c = 3$.

3 $D = \dfrac{v^2}{2a}$, when $v = 6$ and $a = 2$.

4 $s = \dfrac{a + b + c}{2}$, when $a = 7.4$, $b = 3.9$ and $c = 5.8$.

5 $F = \dfrac{mv - mu}{t}$, when $m = 20$, $u = 2$, $v = 5$ and $t = 2$.

6 $L = \dfrac{gt^2}{4\pi}$, when $g = 10$, $t = 1.6$ and $\pi = 3.14$.

7 $A = 2(lb + bw + lw)$, when $l = 20$, $b = 8.5$ and $w = 5$.

8 $s = ut + \dfrac{at^2}{2}$, when $u = 0$, $t = 6$ and $a = 10$.

9 $B = c + d(e - f)$, when $c = 50$, $d = 5$, $e = 16$ and $f = 6$.

10 $B = c + d(e - f)$, when $c = 50.3$, $d = 4.96$, $e = 16.42$ and $f = 5.83$.

Equations

Example 6

I wish to move some bricks, each weighing 3 kg, in my wheelbarrow which weighs 11 kg. The formula for the weight (W kg) that I will be pushing is $W = 3n + 11$, where n is the number of bricks in the wheelbarrow.

(*a*) If there are 8 bricks in the barrow, what weight will I be pushing?

(i) Substituting: $W = (3 \times 8) + 11$.
(ii) Evaluating: $W = 24 + 11 = 35$ kg.

(*b*) If I can comfortably push 50 kg, what is the largest number of bricks that I could put in the barrow and still push it comfortably?

By trying various numbers of bricks, you could eventually come to the answer of 13 bricks, as $(3 \times 13) + 11 = 39 + 11 = 50$.

Can we find a more logical method of tackling this sort of problem, which will work with more awkward numbers as well as with easy ones?

The problem reduces to trying to find n, the number of bricks, when the weight I can push, W, is 50 kg. The formula will become $50 = 3n + 11$, or, writing it the other way round, $3n + 11 = 50$.

This is now an **equation**. It is made up of an 'equals' sign, some numbers and an unknown value (n in this case).

Solving equations: addition

Let us start with a very simple equation, so that we can follow the principles needed to solve the equation.

Example 7

Solve the equation $x + 5 = 8$.

It is easy to see that x must be 3 for the equation to balance. In fact we can imagine that the equation is a balance, with weights of x kg and 5 kg on one scale pan, and a weight of 8 kg on the other.

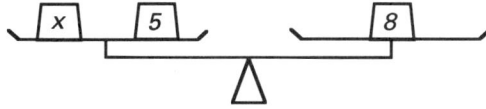

Taking 5 kg from each side will not upset the balance, and will leave $x = 3$ kg.

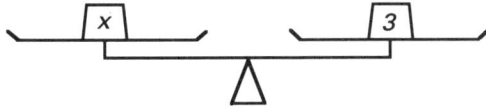

Example 8

Solve the equation $2b + 3 = 13$.

If b is the weight of a box, then we can draw a scale pan diagram, with $2b + 3$ on one side and 13 on the other.

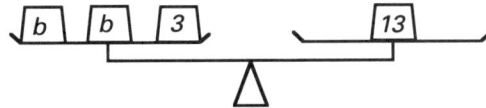

Taking 3 from each side will give

$$2b = 13 - 3$$
$$= 10$$

If $2b = 10$, then

$$b = \frac{10}{2} = 5$$

(Check: $2b + 3 = (2 \times 5) + 3 = 10 + 3 = 13$.)

Example 9

Solve the equation $2h + 7 = 23$.

Draw the scale pan diagram.

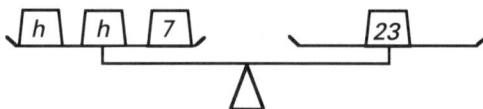

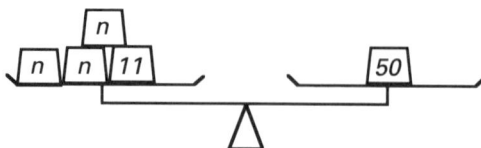

Taking 7 from each side leaves $2h = 16$.

Now dividing by 2 gives $h = 8$.

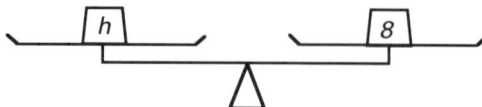

(Check: $2h + 7 = (2 \times 8) + 7 = 16 + 7 = 23$.)

Example 10

Solve the equation $3n + 11 = 50$ (the bricks and wheelbarrow problem).

Draw the scale pan diagram.

Taking 11 from each side leaves

$$3n = 50 - 11$$
$$= 39$$

Dividing by 3 gives $n = \dfrac{39}{3} = 13$.

(Check: $3n + 11 = (3 \times 13) + 11 = 39 + 11 = 50$.)

EXERCISE 4.3

Solve these equations using the scale pan method.

1 $d + 2 = 7$ **2** $s + 4 = 15$

3 $x + 99 = 101$ **4** $20 = c + 7$

5 $2w + 7 = 27$ **6** $3a + 4 = 25$

7 $6 + 2p = 11$ **8** $23 + 4y = 27$

9 $36 = 6 + 10n$ **10** $5t + 16 = 46$

Solving equations: subtraction

Although we cannot put a weight of '−3 kg' on to a scale pan, we can use the same principle to solve equations where a number has been taken away instead of added.

Example 11

Solve the equation $4x - 3 = 17$.

Imagine a weight of '−3 kg'.

In order to cancel it out, we need to *add* 3 kg to each pan; this will change $4x - 3 = 17$ into $4x = 17 + 3 = 20$.

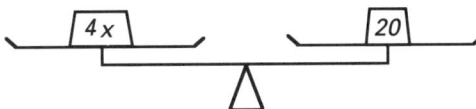

Then, from $4x = 20$, dividing by 4 gives $x = \dfrac{20}{4} = 5$.

(Check: $(4 \times 5) - 3 = 20 - 3 = 17$.)

Example 12

Solve the equation $13 = 2t - 9$.

Draw the scale pans. It doesn't matter into which pan you put 13 as long as the $2t - 9$ goes on the other one!

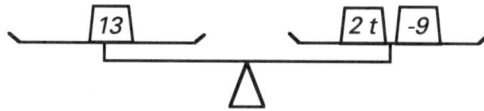

So we can write $2t - 9 = 13$

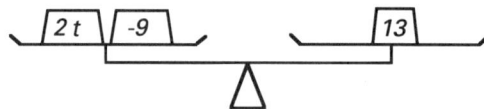

Adding 9 to each pan gives $2t = 13 + 9 = 22$.

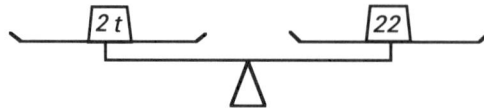

So $t = \dfrac{22}{2} = 11$.

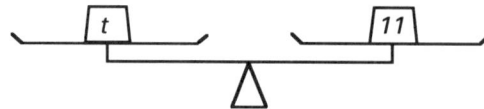

(Check: $(2 \times 11) - 9 = 22 - 9 = 13$.)

EXERCISE 4.4

Solve these equations using the scale pan method.

1 $r - 4 = 5$	**2** $x - 5 = 2$
3 $12 = k - 5$	**4** $b - 6 = 6$
5 $3x - 7 = 23$	**6** $2c - 1 = 15$
7 $4 = 4t - 4$	**8** $6y - 8 = 40$
9 $2a - 3 = 0$	**10** $5d - 4 = 496$

We can extend the principle to three further situations which commonly occur.

Solving equations: fractions

Example 13

Solve the equation $\dfrac{n}{4} = 6$.

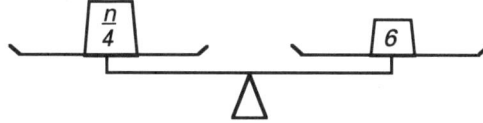

$\dfrac{n}{4}$ means a quarter of n (or $\dfrac{1}{4} \times n$).

To work out n we need to multiply by 4.

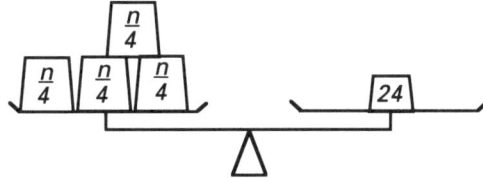

So $\dfrac{n}{4} = 6$ becomes $n = 6 \times 4 = 24$.

(Check: $\dfrac{24}{4} = 6$.)

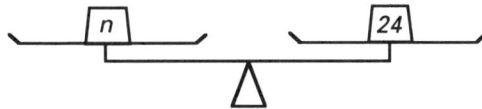

Example 14

Solve the equation $\dfrac{x}{3} + 11 = 15$.

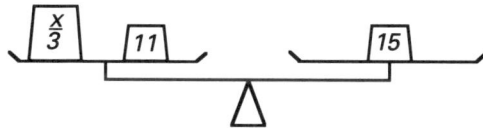

Subtracting 11 from both pans leaves $\dfrac{x}{3} = 4$.

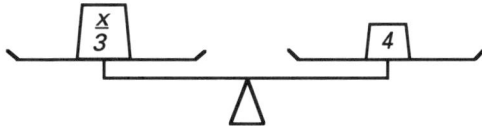

So $x = 4 \times 3 = 12$.

(Check: $\dfrac{12}{3} + 11 = 4 + 11 = 15$.)

Example 15

Solve the equation $\dfrac{2x}{5} - 3 = 5$.

Adding 3 gives $\dfrac{2x}{5} = 8$.

Multiplying by 5: $2x - 3 = 1 \times 5 = 5$.
Adding 3: $2x = 5 + 3 = 8$.
Dividing by 2: $x = 4$.

(Check: $\dfrac{(2 \times 4) - 3}{5} = \dfrac{5}{5} = 1$.)

EXERCISE 4.5

Solve these equations. If you wish, you can omit the scale pan diagrams.

1 $\dfrac{W}{2} = 6$

2 $\dfrac{x}{4} = 10$

3 $\dfrac{m}{7} = 7$

4 $\dfrac{A}{3} = 1$

5 $\dfrac{2c}{3} = 4$

6 $\dfrac{3t}{5} = 6$

7 $\dfrac{v}{2} - 7 = 11$

8 $\dfrac{k}{10} + 5 = 9$

9 $\dfrac{3c}{5} + 8 = 20$

10 $\dfrac{4x}{3} - 4 = 0$

Solving equations: brackets

Example 16

Solve the equation $3(2 + x) = 12$.

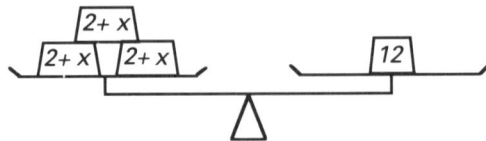

Remember: $3(2 + x)$ means 3 times $(2 + x)$.

Dividing by 3: $2 + x = \dfrac{12}{3} = 4.$

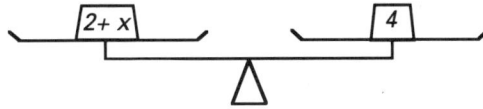

Subtracting 2: $x = 4 - 2 = 2.$

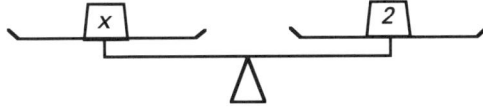

(Check: $3(2 + 2) = 3 \times 4 = 12.$)

Example 17

Solve the equation $\dfrac{2x - 3}{5} = 1.$

Multiplying by 5: $2x - 3 = 1 \times 5 = 5.$
Adding 3: $2x = 5 + 3 = 8.$
Dividing by 2: $x = 4.$

(Check: $\dfrac{(2 \times 4) - 3}{5} = \dfrac{5}{5} = 1.$)

EXERCISE 4.6

Solve these equations.

1 $2(c + 3) = 14$ **2** $3(5 + v) = 18$

3 $\dfrac{m - 4}{3} = 4$ **4** $\dfrac{s - 1}{5} = 8$

5 $6(d + 8) = 60$ **6** $8(z - 8) = 8$

7 $2(3x + 1) = 32$ **8** $4(2t - 14) = 24$

9 $5(2a + 1) = 125$ **10** $3(4s + 1) = 33$

Solving equations: letters occurring on both sides

Example 18

Solve the equation $2x + 7 = x + 13$.

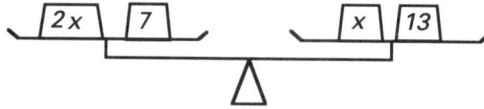

Using the scale pan method, we can make the diagram into a familiar one if we have the x's on one pan only. To do this take an x from each side which leaves

$$x + 7 = 13$$

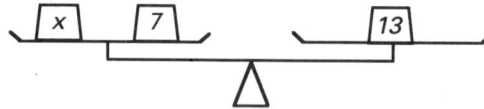

which easily leads to $x = 6$.

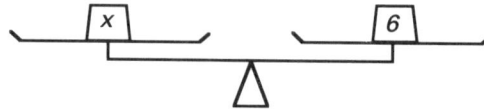

(Check: $(2 \times 6) + 7 = 12 + 7 = 19; 6 + 13 = 19$.)

Example 19

Solve the equation $2x + 5 = 5x - 4$.

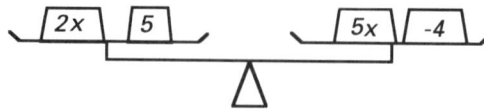

Taking $2x$ from each side leaves

$$5 = 3x - 4$$

Exchanging the pans will give

$$3x - 4 = 5$$

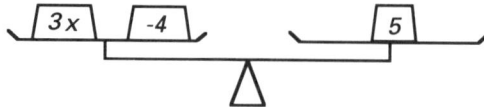

Adding 4: $\quad 3x = 9$

Dividing by 3: $x = 3$

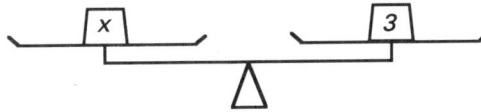

(Check: $(2 \times 3) + 5 = 6 + 5 = 11; (5 \times 3) - 4 = 15 - 4 = 11$.)

EXERCISE 4.7

Solve these equations. The value of x is different in each question.

1 $2x + 1 = x + 6$

2 $4x + 2 = x + 5$

3 $6x - 3 = 4x + 3$

4 $2x + 7 = 3x - 3$

5 $x + 12 = 4x$

6 $5x - 12 = 4x - 3$

7 $8x - 8 = 5x + 10$

8 $9x = 14x - 10$

9 $x + 8 = 5x + 7$

10 $22x + 3 = 20x + 6$

Solving equations: subtraction of the unknown value

Example 20

Solve the equation $15 - 2t = 7$.

Whenever you have an equation with a negative sign in front of the unknown value (in this case, $-2t$), the first step is to *make the sign positive*, by adding (in this case, add $2t$ to each side).

Add $2t$: $15 = 2t + 7$

Taking 7 from each side: $15 - 7 = 2t$
$$8 = 2t$$
$$\text{or} \quad 2t = 8$$

Dividing by 2: $t = 4$

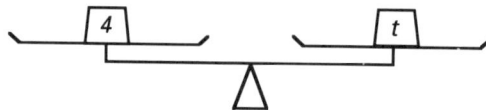

(Check: $15 - (2 \times 4) = 15 - 8 = 7$.)

Example 21

Solve the equation $2q + 9 = 30 - q$.

As we have a $-q$, we need to add q to each side.

This gives
$$2q + 9 + q = 30$$
$$\text{or} \quad 3q + 9 = 30$$

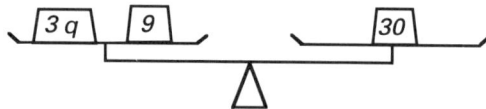

Taking 9 from each side: $3q = 21$

Dividing by 3: $q = 7$

(Check: $(2 \times 7) + 9 = 14 + 9 = 23; \; 30 - 7 = 23$.)

EXERCISE 4.8

Solve these equations.

1 $12 - c = 8$

2 $13 = 18 - x$

3 $3p + 6 = 21 - 2p$

4 $29 - 2y = 5$

5 $20 - 3g = g - 4$

6 $5 + 4b = 11 - 2b$

7 $3(12 - d) = 21$

8 $5(20 - 3z) = 10$

9 $\dfrac{18 - s}{4} = 2$

10 $\dfrac{30 - 5a}{2} = 10$

Transforming formulas

Going back to the bricks and wheelbarrow example, we had
$W = 3n + 11$. So if $n = 8$, then $W = (3 \times 8) + 11 = 24 + 11 = 35$.
We also worked out, by solving the equation $50 = 3n + 11$, that $n = 13$.
How can we rearrange the formula $W = 3n + 11$, to give us

$\quad n = $ (some formula with W in it)?

We can use the scale pan method to help.

First, draw the scale pan diagram.

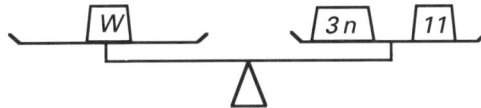

Take 11 from each side, to give $W - 11 = 3n$.

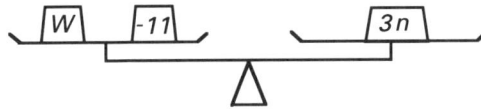

Now divide by 3, to give $n = \dfrac{W - 11}{3}$ or $\frac{1}{3}(W - 11)$.

We have therefore *transformed* the formula from $W = 3n + 11$ to
$n = $ (some formula with W in it). n is now the **subject** of the formula.

$\quad n = \frac{1}{3}(W - 11)$

Example 22

Make k the subject of the formula $p = 5k - 7$.

Add 7: $p + 7 = 5k$

Divide by 5: $\dfrac{p+7}{5}=k$ or $k=\dfrac{p+7}{5}$.

You can check your working by:
  (i) choosing a number for k (say 3),
  (ii) substituting it into the original equation and evaluating p
    ($p=(5\times 3)-7=8$),
  (iii) checking to see that your transformed formula works with these
    numbers ($k=\dfrac{8+7}{5}=\dfrac{15}{5}=3$).

Example 23

Rearrange the formula $s=4-\dfrac{x}{3}$, to make x the subject.

Draw the scale pan diagram.

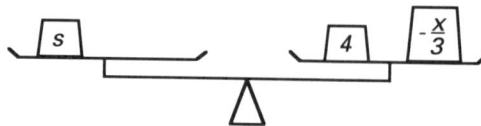

Following the principle established before, make the x term positive by
adding $\dfrac{x}{3}$ to each side: $s+\dfrac{x}{3}=4$

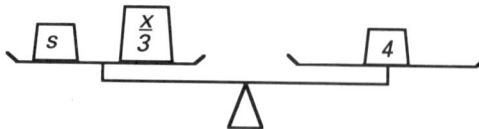

Take s from each side: $\dfrac{x}{3}=4-s$

Multiply by 3: $x=3(4-s)$
Check:
  (i) choose $x=6$, say
  (ii) substitute in the original equation: $s=4-\dfrac{6}{3}=4-2=2$,
  (iii) check: $x=3(4-2)=3\times 2=6$.

EXERCISE **4.9**

Transform these formulas to make x the subject in each case.

1 $t = x + 4$ **2** $m = x - 7$

3 $d = x - 12$ **4** $b = x + 1$

5 $s = 4x$ **6** $l = \dfrac{x}{3}$

7 $a = \dfrac{3x}{5}$ **8** $q = \dfrac{7x}{10}$

9 $c = 2x - 5$ **10** $n = 3x + 4$

11 $k = 6 - x$ **12** $d = 3 - 2x$

13 $w = 3(x + 5)$ **14** $h = \dfrac{2x - 7}{3}$

15 $r = 2(\dfrac{x}{3} - 4)$ **16** $y = mx + c$

17 $t = 5(2x + 3)$ **18** $w = \dfrac{4x - 1}{5}$

19 $2s = 4x - 5$ **20** $d = \dfrac{1}{2}(\dfrac{x}{2} - 7)$

Revision exercises: Chapters 1-4

1 Write the following percentages as fractions, simplifying your answers where possible.

(*a*) 25% (*b*) 40% (*c*) 90% (*d*) 55% (*e*) 15%
(*f*) 43% (*g*) 250% (*h*) 2% (*i*) 84% (*j*) 51%

2 Write the following percentages as decimals:

(*a*) 75% (*b*) 30% (*c*) 23% (*d*) 84%

(*e*) 41% (*f*) 22% (*g*) 99% (*h*) 1%

3 Calculate (*a*) 10% of £45 (*b*) 5% of £6 (*c*) $7\frac{1}{2}$% of £2.50.

4 Find (*a*) 25% of £8.80 (*b*) 2% of £9 (*c*) $3\frac{1}{4}$% of £32.

5 What is the 15% VAT on a radio costing £10.50?

6 Martin achieved an attendance of 95% over a term at school. If there were 120 attendance sessions, how many did he miss?

7 A computer terminal had a fault on a relay which occurred on average 0.5% of the time. Of 1000 messages passed down the relay, how many were expected to be correct?

8 Find the total cost of a cellular telephone advertised at £225 + 15% VAT.

9 Over a year the price of books has risen by $7\frac{1}{2}$%, but the budget of a school's maths department has been cut by 5%. How would you describe the percentage change in purchasing power?

10 What would be the cost of a £340 holiday after the addition of an 8% surcharge?

11 The value of a house rises by $12\frac{1}{2}$% from £34 000. What is its new value?

12 Tyres costing £46 each are sold at a 10% discount. What is the cost of three such tyres after deduction of the discount?

13 A pair of jeans 80cm long shrink by 5% in their first wash. What is their new length?

14 The cost of a £58 railway ticket is to increase by 12%. What will be its new price?

15 The attendance at a football ground falls by 17% from 12 400. What is the new attendance figure?

16 Find the cost of an £80 exhaust system after deduction of a 5% discount, followed by the addition of VAT at 15%.

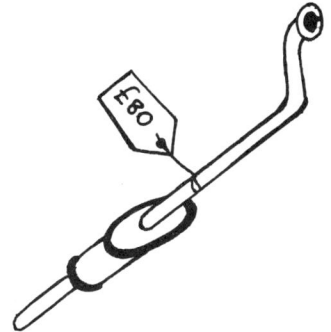

17 Write these fractions as percentages:

(a) $\frac{1}{5}$ (b) $\frac{1}{4}$ (c) $\frac{7}{10}$ (d) $\frac{47}{100}$ (e) $\frac{9}{50}$

(f) $\frac{7}{25}$ (g) $1\frac{1}{2}$ (h) $2\frac{1}{4}$ (i) $\frac{5}{8}$ (j) $\frac{7}{16}$

18 Write these decimals as percentages:

(a) 0.24   (b) 0.36   (c) 0.87   (d) 0.125
(e) 1.5   (f) 4.25   (g) 0.62   (h) 0.01

19 In a class there are 16 girls and 14 boys. (a) What fraction of the class is girls? (b) What fraction is boys? (c) What percentage of the class is girls? (d) What percentage is boys?

20 Write 35p as a percentage of £1.75.

21 There are 12 boys and 18 girls in a class. (a) What percentage of the class is boys? (b) What percentage is girls?

22 The price of a jar of coffee is increased from £1.80 to £1.98. What is the percentage increase?

23 Write 12p as a percentage of £2.40.

24 A computer normally costing £130 has been reduced to £114.40 in a sale. What is the percentage reduction?

25 In a school with 900 pupils there were 783 present during one afternoon. What is the percentage attendance?

26 The price of a bus ticket is increased from 50p to 62p. What is the percentage increase?

27 Out of 40 applicants for a job, 28 were women. What percentage is this?

28 Of 100 cars tested for their MOT, 62 passed. What percentage failed?

29 Of 280 articles sent out by a mail order company, 56 were returned. What percentage were sold?

30 A delivery of sweets costing £9.50 is later sold for a total of £11.21. What is the percentage profit?

31 A number of shoes are bought for £1600, but had to be sold off for £1400. What is the percentage loss?

32 A discount warehouse buys curtain fabric for £2.40 per metre, and expects to make 20% profit. At what price is the fabric sold?

33 A car is bought for £8000 and sold for £6160 the following year. Calculate the percentage loss in value.

34 A ticket to a pop concert was bought for £7.50, and later resold for £9.00. What is the percentage profit?

35 An antique watch was bought at a jumble sale for £12.00, and later sold at auction for £90. What is the percentage profit?

36 Calculate the annual interest on £86 in a bank account which has a rate of interest of 7%.

37 Find the annual interest on £135 in an account which earns a rate of interest of 9%.

38 Calculate the annual interest on £150 in an account which earns a rate of interest of $9\frac{1}{2}$%.

39 What is the interest after 6 months on £450 in an account which earns a rate of interest of 8%?

40 An amount of £200 is kept in an $8\frac{1}{2}$% savings account. How much interest by compound methods will have been earned after 2 years?

41 What will be the compound interest on £300 in an account earning 9% interest per annum for 2 years?

42 What will be the compound interest on £210 in an account earning 10% interest per annum for 3 years?

43 What will be the compound interest on £320 in an account earning 8% interest per annum for 2 years?

44 What will be the compound interest on £80 in an account earning 9% interest per annum for 3 years?

45 June has been quoted £160 gross car insurance. What will she have to pay after deduction of a 60% No Claims Discount?

46 David has been quoted £220 gross car insurance. What will he have to pay after deduction of a 40% No Claims Discount?

47 Martin has been quoted £250 gross car insurance. What will he have to pay after deduction of a 50% No Claims Discount?

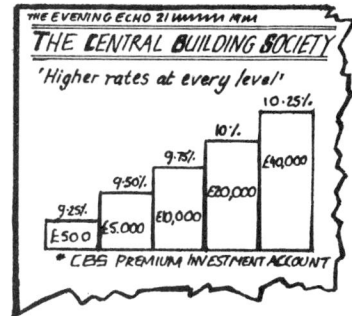

Work out the value of each expression

48 $3x$, when $x = 2$

49 $7k$, when $k = 4$

50 $2y + 3$, when $y = 8$

51 $4a - 7$, when $a = 20$

52 $5t + 6$, when $t = 9$

53 $20 - 2m$, when $m = 6$

54 $50 + 3b$, when $b = 25$

55 $100 - 6t$, when $t = 6$

56 $\frac{c}{4}$, when $c = 24$

57 $30 - \frac{p}{3}$, when $p = 30$

Simplify these expressions.

58 $x + x + 1$

59 $3a - 2 + a$

60 $5b - 3 + 2b$

61 $3m + 2 - m + 3$

62 $7 - 2y + 1 + 5y$

63 $c + 2 + 3c + 4$

64 $7x + 1 - 3x - 2$

65 $12t - 5u + 5t + 12u$

66 $z - 12 + 6x - 12$

67 $a + 7b - a - 6b$

68 $5 \times 3x$

69 $7t \times 2$

70 $3x \times 2y$

71 $k \times 7m$

72 $a \times 2b \times 3c$

73 $4p \times 2q$

74 $12 \times 10t$

75 $a \times b \times c$

76 $6 \times 8c$

77 $4x \times 5y \times 2z$

78 $x \times x \times x \times x$

79 $2x \times 2x$

80 $3x \times 4xy$

81 $st \times 2st \times t$

82 $abc \times ab \times bc$

83 $4d \times 6f$

84 $50n \times 2$

85 $3h \times 6 \times 2k$

86 $100 \times 3t \times 2u$

Work out the value of each expression.

87 $\dfrac{2p}{3}$, when $p=12$

88 $5x+4y$, when $x=1$, $y=2$

89 $6(q+2)$, when $q=5$

90 $3a+4(a+b)$, when $a=5$, $b=7$

91 $\dfrac{x}{3}+\dfrac{x}{4}$, when $x=24$

92 $8c-\dfrac{8}{c}$, when $c=4$

93 $7-\dfrac{2y}{5}$, when $y=10$

94 $52-\dfrac{z}{3}$, when $z=60$

95 $xy+yz+zx$, when $x=y=z=10$

96 $2ab+3a-4b$, when $a=12$, $b=3$

97 $2x^2$, when $x=7$

98 p^2-3p, when $p=4$

99 $4a^2+5b^2$, when $a=2$, $b=1$

100 $3(t^2-3t+5)$, when $t=5$

101 $3b^2-2c^2$, when $b=4$, $c=2$

102 $ut+5t^2$, when $u=20$, $t=3$

103 $6p-p^2+2p^2$, when $p=5$

104 $k(2a^2-3b^2)$, when $k=3$, $a=8$, $t=4$

105 $s(3s+7)$, when $s=6$

106 $\dfrac{1}{3}t^2+\dfrac{1}{4}t$, when $t=12$

Simplify these expressions:

107 $\dfrac{t^2}{t}$

108 $\dfrac{6x^3}{2x}$

109 $2x^2+3x-x^2-x$

110 $12+5b+4-3b$

111 $3(x+3)-2x-9$

112 $\dfrac{12t^2u^3}{3tu^3}$

113 $2(7m^2+3m)-4(2m^2-5m)$

114 $2k(k+1)+k(3k-1)$

115 $\dfrac{30a^3b^2c^3}{3abc^2}$

116 $\dfrac{3x^2}{x}+\dfrac{4x^3}{x^2}-\dfrac{5x^4}{x^3}$

Factorise these expressions:

117 $2x-6$

118 $4a^2+6a$

119 $7t-21u$

120 $12p^2-4p$

121 $ut+\tfrac{1}{2}at^2$

122 $12x^2+6xy$

123 $100-25b$

124 $3pq^2-4p^2q$

125 $12y^2z+8yz^2$

126 $a-6ab$

Evaluate each formula:

127 $k = 3t - 7$ when $(a)\, t = 10$ $(b)\, t = 10.3$.

128 $M = 5x^2 - 8x$ when $(a)\, x = 6$ $(b)\, x = 5.8$.

129 $S = \frac{n}{2}(a + l)$ when $n = 16$, $a = 3$ and $l = 17$.

130 $V = \frac{1}{3}\pi r^2 h$ when $\pi = 3.14$, $r = 4$ and $h = 6$.

131 $t = a(12 - a)$ when $(a)\, a = 7$ $(b)\, a = 7.3$.

132 $l = \frac{p^2 + q^2}{5}$ when $(a)\, p = 6, q = 8$ $(b)\, p = 5.8, q = 8.1$.

133 $A = \frac{1}{2}lb + lc$ when $l = 15$, $b = 4$ and $c = 6$.

134 $C = 3.7P - \frac{84}{Q}$ when $P = 9$ and $Q = 4$.

135 $P = A\left(1 + \frac{r}{100}\right)^2$ when $A = 200$ and $r = 11$.

Solve these equations (use the scale pan method if you wish).

136 $x + 3 = 5$	**137** $3p = 12$	**138** $7 = t + 4$
139 $4a = 20$	**140** $m + 6 = 8$	**141** $2y = 1$
142 $2s + 1 = 7$	**143** $3b + 4 = 25$	**144** $r + 10 = 30$
145 $6c = 60$	**146** $2x + 3 = 9$	**147** $12 = 5y + 2$
148 $k - 3 = 2$	**149** $t - 2 = 15$	**150** $3x - 1 = 11$
151 $42 = 5m + 7$	**152** $8u - 56 = 0$	**153** $12x = 12$
154 $4d + 5 = 45$	**155** $3c - 8 = 25$	**156** $\frac{x}{3} = 7$
157 $\frac{p}{4} = 1$	**158** $\frac{2t}{5} = 4$	**159** $\frac{a}{2} = 7$
160 $\frac{3b}{4} = 6$	**161** $\frac{n}{2} + 3 = 8$	**162** $\frac{w}{5} - 7 = 2$
163 $12 + \frac{2m}{3} = 28$	**164** $\frac{x}{10} + 8 = 18$	**165** $\frac{3y}{5} - 5 = 19$

166 $3(x+1)=12$ **167** $2(t-3)=14$ **168** $4(5+l)=28$

169 $10(m-8)=120$ **170** $3(k-4)=30$ **171** $2(3c+4)=38$

172 $5(2t+5)=65$ **173** $7(a+2)=21$ **174** $4(6p-1)=44$

175 $10(3d-2)=100$ **176** $3x+4=2x+6$ **177** $x+6=2x+1$

178 $h+13=4h+1$ **179** $6s=4s+10$ **180** $3m+5=4m+4$

181 $x+4=2x-3$ **182** $2x+9=5x-3$ **183** $3x=5x-2$

184 $7y-1=4y+20$ **185** $4z-1=2z+9$ **186** $7-x=3$

187 $5+x=12$ **188** $12-2t=4$ **189** $20-3m=2$

190 $4(5-b)=16$ **191** $5(10-3a)=5$ **192** $3(2a-3)=51$

193 $2(25-4c)=26$ **194** $18=3(14-k)$ **195** $7(5-2t)=0$

In each case make the letter shown into the subject of the formula.

196 $A=l\times b$:b   **197** $y=\dfrac{k}{x}$:k   **198** $x=a+l$:l

199 $2t=3x$:x   **200** $m=2n+3$:n   **201** $y=mx+c$:x

202 $t=\dfrac{3}{m}$:m   **203** $c=2\pi r$:r   **204** $l=2(x+5)$:x

5. Communication: numbers (1)

Calculators make arithmetic easier for us to handle, but we still need to understand how to use the calculator properly, and to understand how we can use it to work out the answers to a problem. In Book 3X, Chapter 4, we looked at the importance of brackets and the order in which we carry out numerical operations, particularly when using a calculator.

Order of operations

Example 1

$$11 + 7 \times 3$$

$$11 + 7 \times 3 = 11 + 21 \quad \text{since we do the multiplication first}$$
$$= 32$$

We should really write this problem using brackets to remind us to work out the 7×3 first : $11 + (7 \times 3)$.

Example 2

$$4 \times 8 + 6$$

$$4 \times 8 + 6 = 32 + 6 \quad \text{since again the multiplication is done first}$$
$$= 38$$

We can write this as $(4 \times 8) + 6$ but it is not necessary in this case since the 4×8 appears first in the problem.

Multiplication and division are completed first in a problem, followed by addition and subtraction.

EXERCISE 5.1

Rewrite each of these problems, putting in brackets where
necessary, and work out the answer to the problem.

1 $3 \times 6 + 7$	**2** $17 - 6 + 2$	**3** $8 - 2 \times 4$	**4** $5 - 1 - 2$
5 $18 \div 2 + 4$	**6** $1 + 3 \times 8$	**7** $12 \div 4 + 4$	**8** $16 - 6 \div 2$
9 $16 - 5 + 1$	**10** $8 - 4 - 1$	**11** $14 + 8 \div 2$	**12** $11 + 8 \times 4$
13 $25 - 3 \times 7$	**14** $3 \times 2 \times 4$	**15** $11 - 4 - 1$	**16** $12 \div 3 \times 2$
17 $12 - 7 + 2$	**18** $6 - 3 \times 2$	**19** $18 \div 3 + 2$	**20** $16 - 2 \times 6$
21 $30 \div 2 \times 5$	**22** $2 \times 4 + 6 \div 2$	**23** $16 \div 4 - 12 \div 6$	**24** $5 \times 5 - 2 \times 2$
25 $3 \times 7 + 12 \div 4$	**26** $16 - 9 \div 3 + 1$	**27** $5 \times 4 + 2 \times 3$	**28** $23 \times 2 - 60 \div 3$
29 $9 + 2 \times 3 - 7$	**30** $20 - 6 \times 3 - 2$		

Brackets are not always necessary as the order in which we carry out the
operation may not affect the answer.

$$(9 - 3) \times 2 = 12 \;\Big\}$$
$$9 - (3 \times 2) = 3 \;\Big\}$$

The order is important here since we get
different answers; brackets are needed.

Investigation A

Take the numbers 9, 3 and 2. Write down different combinations of the
numbers using the operations $+, -, \times$ and $\div$ to find out which ones need
brackets. Some examples are shown below, but we can find many
different ones to try.

$(9 \div 3) \times 2$	$9 \div (3 \times 2)$	$(9 - 3) - 2$	$9 - (3 - 2)$
$(9 \times 3) \times 2$	$9 \times (3 \times 2)$	$(9 + 3) \div 2$	$9 + (3 \div 2)$

You should find that some combinations of these operations need
brackets, while others do not.

For example, $9 + 3 + 2$ have repetitive operations and do not need
brackets. They can be worked out easily on a calculator, but we should
understand how calculations like this are done.

Investigation B

Continue Investigation A using different numbers and different
operations. Try extending the problem by using 4 or 5 numbers. Can you
arrive at a set of clearly defined rules to help you decide when brackets
would be necessary in a problem?

Repetitive operations

Example 3

$$
\begin{aligned}
&\ \ \ 5+\ 5+\ 5+\ 5 \\
&=\quad\ \ 10+\ 5+\ 5 \\
&=\qquad\quad 15+\ 5 \\
&=\qquad\qquad\quad 20
\end{aligned}
\qquad
\begin{aligned}
&\ \ \ 5\times\ 5\times\ \ 5\times\ \ 5 \\
&=\quad\ \ 25\times\ \ 5\times\ \ 5 \\
&=\qquad\quad 125\times\ \ 5 \\
&=\qquad\qquad\qquad 625
\end{aligned}
$$

EXERCISE **5.2**

Calculate:

1 $4\times3\times4$ **2** $5+3+7+9$ **3** $9\times8\times7\times2$

4 $4\times1\times3\times7$ **5** $4+9+2+14$ **6** $10\times10\times10\times10$

7 $5\times2\times4\times3\times8$ **8** $1\times2\times5\times3\times2\times7$ **9** $9+4+5+7+8+2$

10 $6\times5\times4\times3\times8\times2$ **11** $5+4+7+2+3$ **12** $7\times5\times4\times2\times1\times3\times6$

$$+\ +\ +\ +\ +$$
$$\times\ \times\ \times\ \times\ \times$$

Positive indices

When we have to work out strings of numbers multiplied together such as those in the above exercise there is a shorter way to write them using an **index**.

$$5\times5\times5\times5=5^4 \quad \text{where 4 is the index.}$$

This means 4 fives multiplied together. We say this is 'five to the power four'. In the same way, $6^5=6\times6\times6\times6\times6=7776$ or '6 to the power 5'.

EXERCISE 5.3

Work out:

1 2^3	**2** 3^3	**3** 4^4	**4** 3^4	**5** 2^5
6 1^4	**7** 10^3	**8** 5^3	**9** 0^2	**10** 9^3
11 8^2	**12** 4^6	**13** 5^5	**14** 3^6	**15** 12^3
16 7^4	**17** 11^3	**18** 13^2	**19** 15^3	**20** 2^{10}

Find the missing index.

21 $2^? = 8$	**22** $2^? = 2$	**23** $2^? = 16$	**24** $2^? = 32$	**25** $3^? = 27$
26 $4^? = 64$	**27** $8^? = 64$	**28** $5^? = 125$	**29** $6^? = 216$	**30** $9^? = 729$

Example 4

$$3^2 \times 4^3 = 3 \times 3 \quad \times \quad 4 \times 4 \times 4 = 9 \times 64 = 576$$

EXERCISE 5.4

Work out:

1 $2^2 \times 2^3$	**2** $3^3 \times 2^2$	**3** $2^4 \times 4^2$	**4** $5^2 \times 3^3$	**5** $2^7 \div 2^5$
6 $4^3 \div 2^2$	**7** $2^8 \div 2^2$	**8** $3^5 \div 9^2$	**9** $8^2 \times 6^3$	**10** $5^4 \times 4^3$
11 $6^4 \times 3^3$	**12** $5^4 \div 5^2$	**13** $4^5 \times 3^4$	**14** $8^3 \div 4^4$	**15** $9^3 \div 3^4$
16 $6^3 \times 8^4$				

Standard form notation

Using your calculator, work out 9^9.
Depending on your calculator, you should have found either an E
appearing on the display, or a number which looks like 3.8742 08. If you
have a calculator with a large display you might have the complete answer
shown: 387420489.

As a calculator display usually shows a maximum of eight digits, most
calculators cannot show this number as it is nine digits long. What does
the calculator do when it tries to show the answer?

Some calculators show an E for error, indicating that it cannot show
the number as it is too large. This is called **overflow**, and the answer to the
problem cannot be found. Scientific calculators use **standard form** to
show the number. This is a method of shortening larger numbers in a
standard form.

$$3.8742 \ 08 \quad \text{means} \quad 3.8742 \times 10^8$$
$$\text{or } 3.8742 \times 10 \times 10 \times 10 \times 10 \times 10 \times 10 \times 10 \times 10$$

We have to multiply the decimal by 10 eight times to show it as an
ordinary number. Remember that each time we multiply by ten the
decimal moves one place to the right, so when we multiply by 10^8 the
decimal moves up 8 places to the right: $3.8742 \times 10^8 = 387\ 420\ 000$,
which is an approximation to the complete answer shown above. In
standard form we therefore have two parts to the number. The first part is
a decimal with the decimal point after the first digit. The second part is
the power of ten by which we must multiply the decimal to change it back
into an ordinary number.

Example 5

Write as ordinary numbers (a) 4.271×10^5 (b) 2.04 06,
as shown on a calculator.

(a) $4.271 \times 10^5 = 427100. = 427100$
(b) $2.04 \ 06 = 2.04 \times 10^6 = 2040000.$

EXERCISE 5.5

Write as ordinary numbers.

1 8.324×10^2 **2** 4.2714×10^3 **3** 5.21×10^2 **4** 2.01×10^3

5 1.29×10^4 **6** 1.899×10^5 **7** 4.1×10^5 **8** 4.78×10^4

9 9.121×10^6 **10** 8.905×10^5 **11** 7.12×10^4 **12** 9.1×10^7

13 4.57×10^8 **14** 5.55×10^7 **15** 1.01×10^6 **16** 9.73×10^9

Write the following numbers, shown on a calculator display, as ordinary numbers.

17 1.21 04   **18** 8.7 03   **19** 2.91 05   **20** 4.93 03

21 8.2 06   **22** 4.8 08   **23** 7.613 06   **24** 2.91 09

25 4.97 08   **26** 4.02 07   **27** 3.976 05   **28** 1.0045 04

29 7.765 02   **30** 9.65 10

Writing ordinary numbers in standard form can also be done in a similar way, working backwards. To write the number 51 000 in standard form we first insert the decimal point behind the first digit: 5.1000. How many places have we moved the decimal point? Four places, because the decimal point is always at the end of a whole number. Removing the trailing zeros gives us:

$$51\,000 = 5.1 \times 10^4$$

This is useful because we can enter large numbers into scientific calculators in standard form. First put in the decimal 5.1, then press the *exponential* button, which is usually marked as EXP, or E, or EE. This gives you the space in the display to enter the index 4. The display should now read: 5.1 04.

Example 6

Write 342 000 in standard form.

342 000 becomes 3.420 00, and we have moved the decimal point five places. So $342\,000 = 3.42 \times 10^5$.

EXERCISE 5.6

Write each number in standard form.

1 7560	**2** 423 000	**3** 5 270 000	**4** 42.1
5 92.7	**6** 4380	**7** 45 000	**8** 45 000 000
9 432.8	**10** 40 050	**11** 3000	**12** 3532
13 79 000	**14** 2037	**15** 2 798 000	**16** 117
17 4 300 000	**18** 3052	**19** 604 000	**20** 932 170
21 5423.1	**22** 235.12	**23** 42.936	**24** 57 420.1
25 982.77	**26** 34.78	**27** 909.09	**28** 3 256 701
29 3000.03	**30** 100 000 000		

Negative indices and standard form

$$52000 = 5.2 \times 10^4$$
$$5200 = 5.2 \times 10^3$$
$$520 = 5.2 \times 10^2$$
$$52 = 5.2 \times 10^1$$
$$5.2 = 5.2 \quad \text{already in standard form}$$
$$0.52 = 5.2 \times 10^?$$
$$0.052 = 5.2 \times 10^?$$

This list of numbers shows their standard forms as they are reduced each time by a factor of 10. Can you complete the indices for the last two? Looking at the pattern of indices we reduce by one each step. 4, 3, 2, 1, ..., −1, −2 is the obvious choice, so we have:

$$0.52 = 5.2 \times 10^{-1}$$
$$0.052 = 5.2 \times 10^{-2}$$
$$0.0052 = 5.2 \times 10^{-3}$$

This makes sense, but what does a negative index mean?

$0.52 = 5.2 \times 10^{-1}$ the decimal point has moved one place to the *right*.

The negative index moves the decimal point in the opposite direction to when we had a positive index. Also a move of the decimal point to the right tells us we are *dividing* by factors of ten, not multiplying. This is because

$$10^{-1} \text{ means } 1 \div 10 \text{ or } \frac{1}{10}$$

$$10^{-2} \text{ means } 1 \div 100 \text{ or } \frac{1}{100}$$

$$10^{-3} \text{ means } 1 \div 1000 \text{ or } \frac{1}{1000}$$

Example 7

3^{-2} Without the negative sign $3^2 = 9$.

     With the negative sign $3^{-2} = \frac{1}{9}$.

5^{-3} Without the negative sign $5^3 = 125$.

     With the negative sign $5^{-3} = \frac{1}{125}$.

The negative sign has the effect of turning an answer upside down.

EXERCISE 5.7

Work out:

1 2^{-3} $\frac{1}{8}$	**2** 3^{-3}	**3** 8^{-2}	**4** 2^{-5}	**5** 3^1
6 4^0 4	**7** 3^{-4}	**8** 2^{-1}	**9** 1^{-4}	**10** 10^{-3}
11 4^{-2} $\frac{1}{16}$	**12** 6^{-1}	**13** 5^{-2}	**14** 1^0	**15** 12^{-2}
16 7^{-2}	**17** 3^{-3}	**18** 9^0	**19** 8^{-3}	**20** 7^1
21 2^{-4}	**22** 1^{-1}	**23** 3^{-5}	**24** 4^{-4}	**25** 2^0

Example 8

(*a*) Write 3.52×10^{-4} as an ordinary number. (*b*) Write $0.004\,79$ in standard form.

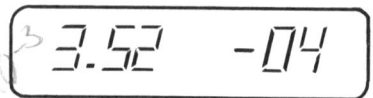

(*a*) $3.52 \times 10^{-4} = 0.000352 = 0.000\,352$

(*b*) $0.004\,79$ becomes 4.79×10^{-3} as we have moved the decimal point 3 places to the left.

EXERCISE 5.8

Write as ordinary numbers:

1 1.21×10^{-2} **2** 1.897×10^{-2} **3** 4.108×10^{-3} **4** 3.95×10^{-4}

5 4.1×10^{-2} **6** 2.9×10^{-3} **7** 3.73×10^{-1} **8** 4.709×10^{-4}

9 2.949×10^{-3} **10** 3.9×10^{-5} **11** 3.7895×10^{-2} **12** 8.72×10^{-1}

13 9.5×10^{-7} **14** 8.904×10^{-6} **15** 2.91×10^{-8} **16** 3.445×10^{-4}

Write in standard form:

17 0.09   **18** 0.071   **19** 0.0084   **20** 0.912

21 0.032   **22** 0.000 54   **23** 0.000 921   **24** 0.000 045

25 0.008 19   **26** 0.082   **27** 0.000 000 18   **28** 0.000 067 1

29 0.000 000 008   **30** 0.009 004   **31** 0.008 008   **32** 0.000 050 109

Investigation C

Using standard form find the greatest and least numbers you could display on your calculator. If possible express these numbers both as standard form numbers and as ordinary numbers.

Negative indices and integers

$$520 = 5.2 \times 10^2$$
$$52 = 5.2 \times 10^1$$
$$5.2 = 5.2 \times 10^?$$
$$0.52 = 5.2 \times 10^{-1}$$
$$0.052 = 5.2 \times 10^{-2}$$

Any number to the power of 1 remains the same. Hence, 10^1 is the same as 10, and 8^1 is the same as 8, etc. The power of 1 does not change the number:

$$52 = 5.2 \times 10^1$$

If we attempt to complete the pattern of powers in the diagram it would seem logical to say that $5.2 = 5.2 \times 10^0$. What is 10^0? As $5.2 = 5.2 \times 10^0 = 5.2 \times 1$, then 10^0 is the same as 1.

Indeed, any number to the power of 0 is just 1. So $2^0 = 1$, $4^0 = 1$, $9^0 = 1$, $100^0 = 1$. Anything to the power of 0 is always 1. Is $0^0 = 1$?

In the work we have already done we found 10^{-2} was $1 \div 10^2$, or $\dfrac{1}{100}$.

A negative index was discovered to be the reciprocal, to mean 'one over'. This applies to any number.

Investigation D: extended course-work investigation into indices

1 $2^2 \times 2^3 = 4 \times 8 = 32 = 2^5$

In this problem the answer can be converted into a power of 2. Will this work with two 2s raised to any combination of indices? Try various combinations of powers of numbers to find out.

2 By creating your own similar problems using only one base number raised to different powers (e.g. $3^7 \times 3^4 \times 3^{-2}$), we should be able to find a short-cut method to find the answer in terms of a power:

$$3^2 \times 3^4 = 3^? \qquad 2^3 \times 2^5 = 2^?$$

Try different combinations of powers, always using the same base number. Continue the investigation until you can arrive at a rule to help you find the answer quickly. Does the same rule apply when there are three or four numbers like $2^2 \times 2^3 \times 2^2 \times 2^4$?

3 Repeat the investigation carried out in part 2 above, but use division instead of multiplication: $2^6 \div 2^4 = ?$

Can you find a similar rule for division?

4 Once you are satisfied you have a set of rules for multiplying and dividing numbers to a certain power, test the rules using negative powers to see if they will still work. Write down any calculations you do, and what you find out.

RULE BOOK

Rounding

In Book 3X, Chapter 3, we found it was necessary to round off to the nearest penny when we had answers which contained fractions of a penny.

$£1.25\underline{41} \approx £1.25$ here we rounded *down* since the additional digits were less than $\frac{1}{2}$ penny.

$£0.12\underline{6} \approx £0.13$ we add on one more penny since the last figure was more than $\frac{1}{2}$ penny.

This same principle can be used to help us write down decimals. Occasionally a calculator leaves us with a very long decimal answer. It would be impractical to keep writing down a decimal as long as, say 0.432 746 1. We therefore round it off, in the same way that we did when handling money.

$0.4 \leftarrow$ 1st decimal place $: \dfrac{1}{10}$

$3 \leftarrow$ 2nd decimal place $: \dfrac{1}{100}$

$2 \leftarrow$ 3rd decimal place $: \dfrac{1}{1000}$

$7 \leftarrow$ 4th decimal place $: \dfrac{1}{10000}$

· etc.

Each column behind the decimal can be counted: the $\frac{1}{10}$ digit is called the 1st decimal place, the $\frac{1}{100}$ digit is called the 2nd decimal place, and so on. To correct a number to one decimal place we round off after the first decimal place:

0.4 327461 = 0.4 to 1 decimal place.
and 0.43 27461 = 0.43 to 2 decimal places.

To correct the number to three decimal places we round off after the third decimal place: 0.432 7461 = 0.433 to 3 decimal places. The number is changed to 0.43$\underline{3}$ since the next digit is 7 and we have to round up. Remember, round up if the following digit is 5 or more.

0.4327 461 = 0.4327 to 4 decimal places.
0.43274 61 = 0.43275 to 5 decimal places.

Correcting to a certain number of decimal places also makes an answer sensible. We would find it difficult to draw a line to a greater accuracy than 1 mm, so an answer of 8.427 cm would be corrected to one decimal place: 8.4 cm.

Example 9

Round off 13.239 to (*a*) 1 decimal place (*b*) 2 decimal places.

(*a*) 13.2 39 = 13.2 to 1 d.p.
(*b*) 13.23 9 = 13.24 to 2 d.p.

EXERCISE 5.9

Write down the following numbers rounded to the number of decimal places (d.p.) stated.

1 6.481   (*a*) 2 d.p.   (*b*) 3 d.p.

2 4.9804   (*a*) 2 d.p.   (*b*) 3 d.p.

3 1.8947   (*a*) 3 d.p.   (*b*) 2 d.p.   (*c*) 1 d.p.

4 101.1717   (*a*) 3 d.p.   (*b*) 2 d.p.   (*c*) 1 d.p.

5 0.033   (*a*) 2 d.p.   (*b*) 1 d.p.

6 1.3507   (*a*) 2 d.p.   (*b*) 3 d.p.

7 81.2954   (*a*) 1 d.p.   (*b*) 2 d.p.   (*c*) 3 d.p.

8 1.0077   (*a*) 3 d.p.   (*b*) 2 d.p.

9 8.8099   (*a*) 3 d.p.   (*b*) 2 d.p.

10 28.2939   (*a*) 3 d.p.   (*b*) 2 d.p.   (*c*) 1 d.p.

11 2.90035   (*a*) 3 d.p.   (*b*) 1 d.p.

12 0.00727   (*a*) 3 d.p.   (*b*) 4 d.p.

13 9.9999   (*a*) 1 d.p.   (*b*) 2 d.p.

14 17.079   (*a*) 1 d.p.   (*b*) 2 d.p.

15 0.000355   (*a*) 3 d.p.   (*b*) 4 d.p.   (*c*) 5 d.p.

Significant figures

Another way of rounding off numbers is to use a method called significant figures (s.f.), where we round numbers off to leave the most important or **significant** digits. The most significant figures are those non-zero digits which come first in a number, those which are on the immediate left. This is because these numbers are greatest in value.

Example 10

What are the significant figures of the numbers 21.538 and 0.0906? The first significant figure is the first number which is not zero:

$2 \leftarrow$ 1st s.f.
$1 \leftarrow$ 2nd s.f.
.
$5 \leftarrow$ 3rd s.f.
$3 \leftarrow$ 4th s.f.
$8 \leftarrow$ 5th s.f

0
.
0
$9 \leftarrow$ 1st s.f.
$0 \leftarrow$ 2nd s.f.
$6 \leftarrow$ 3rd s.f.

When we round off after significant figures we do so in the same way as before; if the next digit after the last we want is 5 or more, we round up.

Example 11

Write (*a*) 21.538 to 3 s.f. (*b*) 0.0906 to 2 s.f.

(*a*) $21.538 = 21.5$ to 3 s.f.
    ↓↓↓
    3 s.f.
(*b*) $0.0906 = 0.091$ to 2 s.f., rounding up.
      ↓↓
      2 s.f.

EXERCISE **5.10**

Write the following numbers rounded to the number of significant
figures (s.f.) stated.

1 6.345   (2 s.f.)   **2** 0.079   (1 s.f.)   **3** 6.72   (2 s.f.)

4 0.0054 (1 s.f.)   **5** 0.53   (1 s.f.)   **6** 13.7425 (4 s.f.)

7 0.000 83(1 s.f.)   **8** 3.207   (2 s.f.)   **9** 0.3964 (2 s.f.)

10 3.025   (2 s.f.)   **11** 14.99 (3 s.f.)   **12** 174.25 (3 s.f.)

13 1.0405   (*a*) 2 s.f.   (*b*) 3 s.f.

14 2.01   (*a*) 1 s.f.   (*b*) 2 s.f.

15 0.949   (*a*) 2 s.f.   (*b*) 3 s.f.

16 73.497   (*a*) 3 s.f.   (*b*) 4 s.f.

17 8.8885   (*a*) 1 s.f.   (*b*) 2 s.f.

18 1.0005   (*a*) 4 s.f.   (*b*) 3 s.f.

19 0.000 64   (*a*) 4 s.f.   (*b*) 3 s.f.

20 6.000 08   (*a*) 2 s.f.   (*b*) 5 s.f.

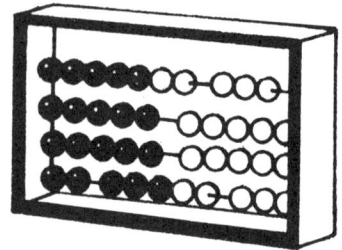

Example 12

Round off the number 5438 to the nearest 100.

$5438 \approx 5400$ since the 4 represents hundreds, we round off to this
number.

This example is similar to those done before in Book 3X, Chapter 3. In
rounding off to the nearest hundred we have also rounded the number
correct to 2 significant figures. When we are dealing with large numbers
we must remember to replace the units or tens we remove in rounding by
zeros, otherwise we would change the value of the number.

Example 13

Write (*a*) 32 430 correct to 2 s.f. (*b*) 47 893 correct to 3 s.f.

(*a*) $32.430 = 32\,000$ correct to 2 s.f.
   ↓↓
   2 s.f.
(*b*) $47\,893 = 47\,900$ correct to 3 s.f.
   ↓↓↓
   3 s.f.

EXERCISE 5.11

Write the following numbers rounded off to the number of significant figures (s.f.) stated.

1 738 (2 s.f.)   *2* 631 (2 s.f.)   *3* 1491 (2 s.f.)

4 582 (1 s.f.)   *5* 4229 (1 s.f.)   *6* 13 842 (3 s.f.)

7 38   (1 s.f.)   *8* 7152 (2 s.f.)   *9* 10 907 (3 s.f.)

10 939   (*a*) 1 s.f.   (*b*) 2 s.f.

11 66 666   (*a*) 2 s.f.   (*b*) 3 s.f.

12 527.42   (*a*) 1 s.f.   (*b*) 2 s.f.

13 497   (*a*) 1 s.f.   (*b*) 2 s.f.

14 613.52   (*a*) 1 s.f.   (*b*) 2 s.f.

15 1183.73   (*a*) 2 s.f.   (*b*) 3 s.f.

16 1045.97   (*a*) 2 s.f.   (*b*) 3 s.f.

17 9394.7   (*a*) 1 s.f.   (*b*) 3 s.f.

18 276.83   (*a*) 1 s.f.   (*b*) 2 s.f.

Reciprocals

When using formulas or performing calculations there are times when we need to turn an expression, decimals or fractions upside down. When we do this we are finding the **reciprocal** of the number. To find a reciprocal we turn the number upside down.

Example 14

Write down the reciprocal of (*a*) 5 (*b*) $\frac{1}{2}$ (*c*) $3\frac{1}{4}$.

(*a*) $5 = \frac{5}{1}$; the reciprocal is therefore $\frac{1}{5}$.

(*b*) Starting with $\frac{1}{2}$, the reciprocal is $\frac{2}{1} = 2$.

(*c*) $3\frac{1}{4} = \frac{13}{4}$, so the reciprocal is $\frac{4}{13}$.

The reciprocals of decimals are decimals themselves. To find the reciprocal of 1.25 we need to calculate $\frac{1}{1.25} = 1 \div 1.25 = 0.8$.

Alternatively, many calculators have a reciprocal button $\frac{`1`}{x}$ which will find reciprocals for us. Entering the decimal 1.25 followed by $\frac{`1`}{x}$ gives the reciprocal 0.8.

Example 15

Calculate the reciprocal of (*a*) 0.5 (*b*) 1.7. Write your answer to 4 decimal places where necessary.

(*a*) 0.5 becomes 2.
(*b*) 1.7 becomes 0.5882 to 4 decimal places.

EXERCISE 5.12

Write down the reciprocals of the following numbers:

1 2	**2** $\frac{1}{4}$	**3** $\frac{3}{8}$	**4** $2\frac{1}{2}$	**5** 5	**6** $\frac{4}{9}$
7 $\frac{3}{5}$	**8** $2\frac{1}{3}$	**9** 4	**10** $1\frac{1}{6}$	**11** $\frac{2}{3}$	**12** $\frac{4}{7}$
13 $8\frac{1}{2}$	**14** $\frac{8}{15}$	**15** $\frac{2}{17}$	**16** $4\frac{2}{3}$	**17** $3\frac{1}{16}$	**18** $\frac{7}{16}$
19 $\frac{7}{8}$	**20** $8\frac{1}{7}$				

Write down the reciprocals of these decimals, writing your answers to 4 decimal places where necessary.

21 0.25	**22** 1.5	**23** 0.4	**24** 0.12	**25** 3.4	**26** 5.7
27 0.27	**28** 0.76	**29** 8.51	**30** 0.23	**31** 10.4	**32** 0.29
33 0.843	**34** 0.461	**35** 14.3			

Investigation E

Find the reciprocal of a number, either a fraction or a decimal, and then find the reciprocal of the reciprocal. What do you find? Try the experiment again with another number. Will it work every time? Can you explain why?

Squares

$$2^2 = 2 \times 2 = 4$$

2^2 is sometimes called 2 'squared' because we carry out the same process when we find the area of a square with a side of length 2 units. Similarly the square of 4 is $4^2 = 4 \times 4 = 16$.

The same applies to decimals: $2.5^2 = 2.5 \times 2.5 = 6.25$.

EXERCISE 5.13

Find the squares of these numbers.

1 3	**2** 11	**3** 5	**4** 15	**5** 24	**6** 35
7 19	**8** 83	**9** 7	**10** 60	**11** 13	**12** 47
13 10	**14** 36	**15** 1.2	**16** 2.7	**17** 8	**18** 9.3
19 0.23	**20** 103	**21** 7.4	**22** 3.9	**23** 0.17	**24** 12
25 14.6	**26** 4.7	**27** 18.1	**28** 127	**29** 8.1	**30** 3.07

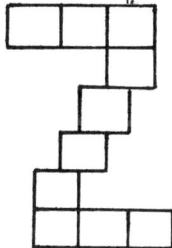

Square roots

Finding a square root means finding the number we started with when we have just worked out a squared number, or doing the opposite of finding a square. 7 'squared' $= 7^2 = 7 \times 7 = 49$ is a square number. The **square root** of 49 is 7, since $49 = 7 \times 7 = 7^2$. We use the symbol $\sqrt{}$ for the square root.

$$\sqrt{49} = 7$$

Similarly $\sqrt{9} = 3$ because $3^2 = 3 \times 3 = 9$.

Most calculators have a square root button. You put in the number, press the button, and the calculator will find the square root for you.

Example 16

Find the square roots of (a) 2.25 (b) 5.2, writing your answers correct to 4 significant figures.

(a) $\sqrt{2.25} = 1.5$
(b) $\sqrt{5.2} = 2.280\ 35 = 2.280$ to 4 s.f.

EXERCISE 5.14

Find the square root of the following numbers.

1 36	**2** 81	**3** 16	**4** 289	**5** 441	**6** 196
7 12.25	**8** 529	**9** 2.56	**10** 324	**11** 0.64	**12** 3.61
13 784	**14** 17.64	**15** 1089	**16** 841	**17** 28.09	**18** 2.89
19 6724	**20** 68.89	**21** 75.69	**22** 8281	**23** 18.49	**24** 50.41
25 5776					

Find the square root of the following numbers writing your answer correct to 4 significant figures.

26 45	**27** 168	**28** 32	**29** 101	**30** 68	**31** 73
32 192	**33** 24.5	**34** 32.7	**35** 0.59		

Pythagoras

Thousands of years ago the Egyptians needed exact right angles in order to construct the corners of buildings accurately.

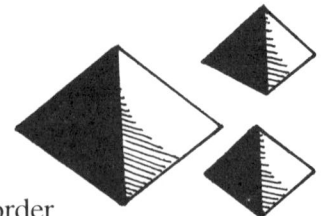

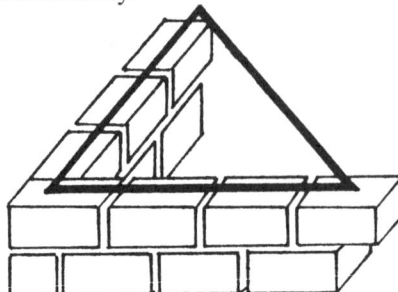

A Greek mathematician called Pythagoras spent some time investigating these right-angled triangles, and a rule (or theorem) is named after him. The most famous of these triangles is this one:

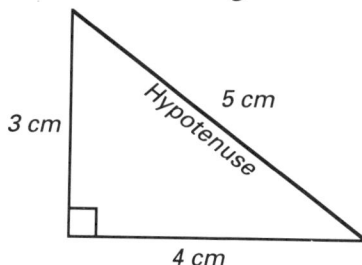

Draw this triangle accurately into your book with ruler and compasses, and check that it is right-angled. Square each of the sides.

$$3^2 = 9$$
$$4^2 = 16$$
$$5^2 = 25$$

What do we find? We have a sum which tells us that

$$3^2 + 4^2 = 5^2$$

or $\quad 9 + 16 = 25$

Pythagoras proved that this *only* works for right-angled triangles. If we find the squares of the lengths of the two shorter sides and add them, this will then give us the square of the length of the longest side, which we call the **hypotenuse**.

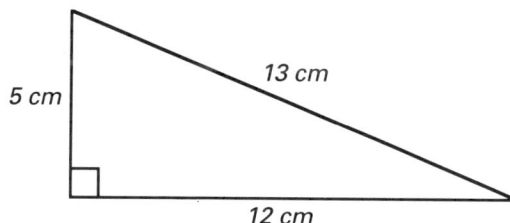

Using ruler and compasses draw this triangle accurately and check it is right-angled by measuring the angles with your protractor. Square all the sides and check that Pythagoras' theorem works on this triangle. It should do if it is right-angled.

Example 17

Is this triangle right-angled?

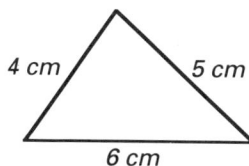

$$4^2 = 16$$
$$5^2 = 25$$
$$6^2 = 36$$

But $16 + 25 \neq 36$ and $4^2 + 5^2 \neq 6^2$.
So Pythagoras' theorem does not work on this triangle, and we can be sure it is not a right-angled triangle.

EXERCISE **5.15**

Use Pythagoras' theorem to find out whether the following triangles
are right-angled or not. They have not been drawn accurately.

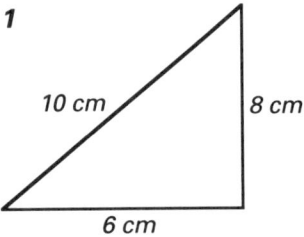

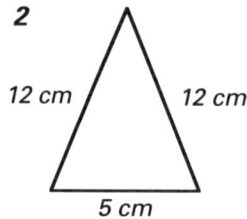

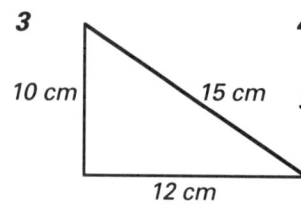

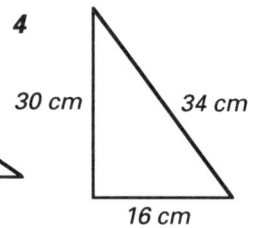

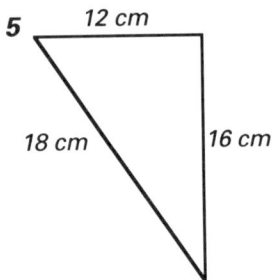

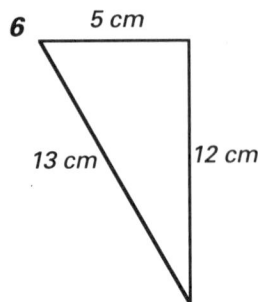

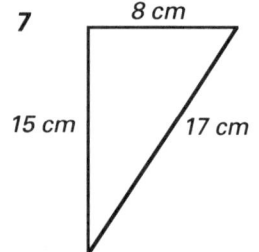

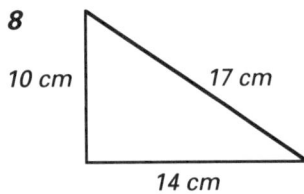

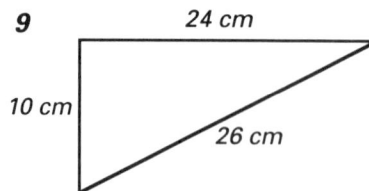

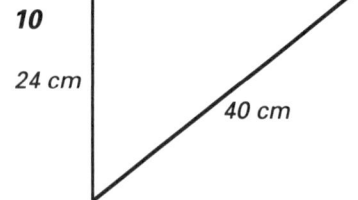

1

10 cm   8 cm

6 cm

2

12 cm   12 cm

5 cm

3

10 cm   15 cm

12 cm

4

30 cm   34 cm

16 cm

5

12 cm

18 cm   16 cm

6

5 cm

13 cm   12 cm

7

8 cm

15 cm   17 cm

8

10 cm   17 cm

14 cm

9

24 cm

10 cm

26 cm

10

32 cm

24 cm

40 cm

Investigation F

Using a long piece of string or rope make a right-angled triangle using the
dimensions of a Pythagorean triangle in metres, clearly marking the
vertices. Either stake out or space out the triangle to help you explain
clearly how this triangle might have been used in the construction of
buildings.

Using Pythagoras' theorem

When we have a right-angled triangle and we know the two shortest sides
we can use Pythagoras' theorem to find the hypotenuse.

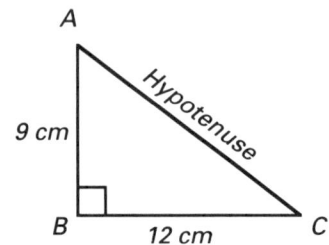

A

9 cm   Hypotenuse

B   12 cm   C

Example 18

Use Pythagoras' theorem to find the length of the hypotenuse.

Using Pythagoras' theorem: $AC^2 = AB^2 + BC^2$
$$AC^2 = 9^2 + 12^2$$
$$AC^2 = 81 + 144$$
$$AC^2 = 225$$

To find AC we now work backwards and find the square root:

$$AC = \sqrt{225} = 15 \text{ cm}$$

The length of the hypotenuse is 15 cm.

EXERCISE 5.16

For each triangle find the length of the hypotenuse, writing your answers correct to 2 decimal places where necessary.

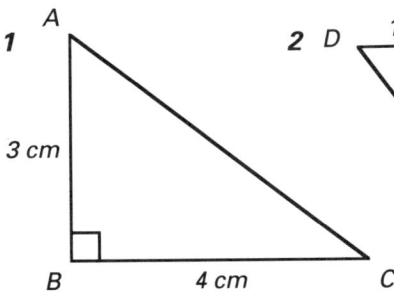

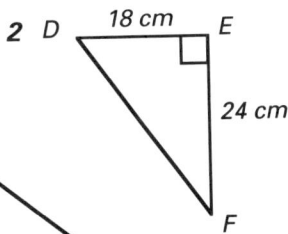

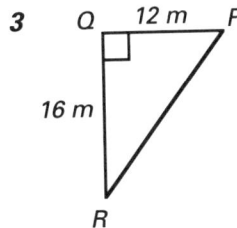

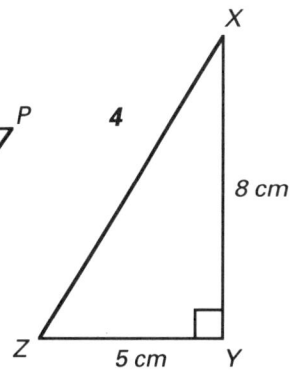

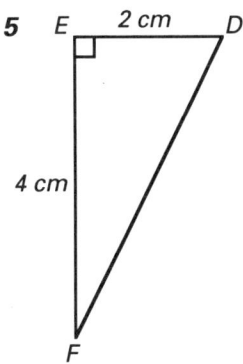

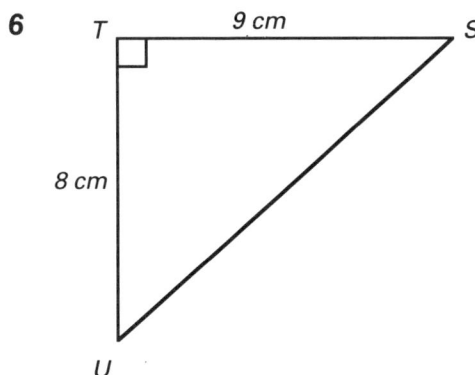

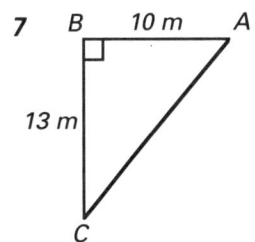

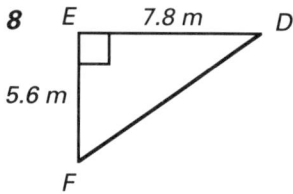

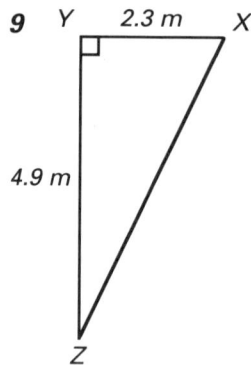

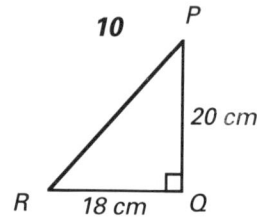

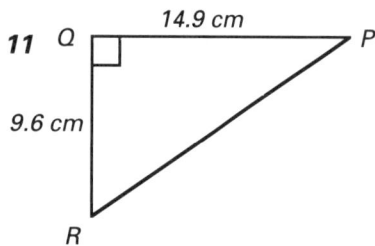

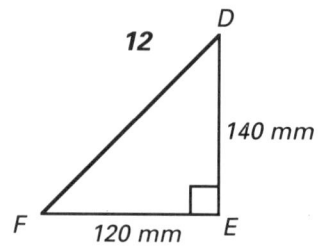

8 E ⌐ 7.8 m D

5.6 m

F

9 Y ⌐ 2.3 m X

4.9 m

Z

10 P

20 cm

R 18 cm Q

11 Q ⌐ 14.9 cm P

9.6 cm

R

12 D

140 mm

F 120 mm E

EXERCISE 5.17

1 Find the length of the diagonal.

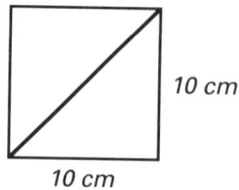

10 cm

10 cm

2 Find the length of the ladder needed to reach up across the corridor.

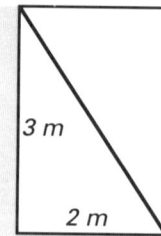

3 m

2 m

35 miles

3 A plane travels 45 miles due North, and then 35 miles due West. How far is the plane from its point of departure?

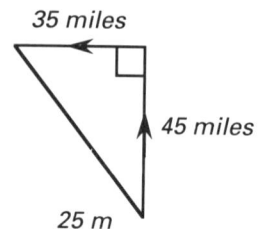

45 miles

25 m

4 A clothes line is to be hung across a garden. How long should it be?

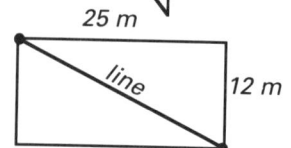

line 12 m

Example 19

Find the length of the side BC.

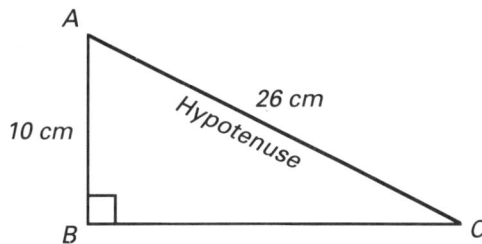

In this example we have already been given the hypotenuse, along with one of the sides. It is one of the shorter sides we have to find. Writing down Pythagoras' theorem in the correct order, starting with the hypotenuse:

$$AC^2 = AB^2 + BC^2$$
$$26^2 = 10^2 + BC^2$$
$$676 = 100 + BC^2$$

So BC^2 is $676 - 100 = 576$
To find BC, we find the square root:

$$BC = \sqrt{576} = 24 \text{ cm}$$

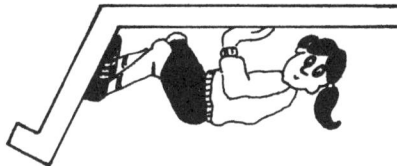

EXERCISE 5.18

For each triangle find the length of the missing side, writing your answers correct to 2 decimal places where necessary.

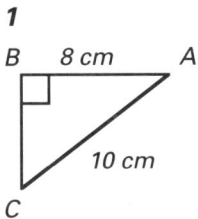

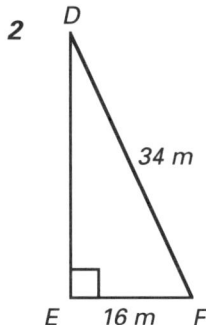

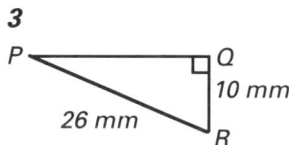

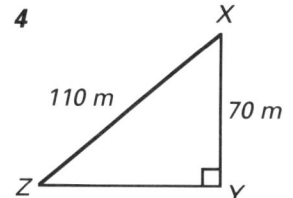

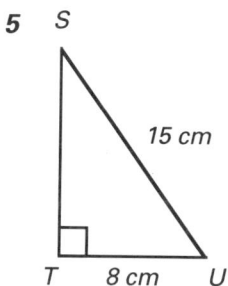

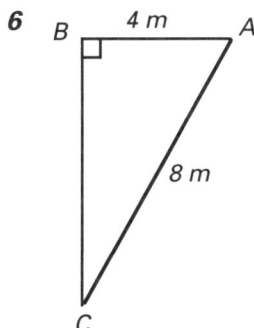

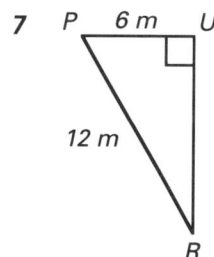

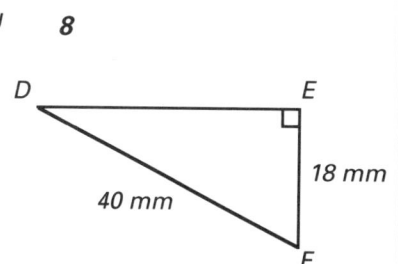

1

B   8 cm   A

10 cm

C

2

D

34 m

E   16 m   F

3

P   Q

10 mm

26 mm

R

4

X

110 m   70 m

Z   Y

5

S

15 cm

T   8 cm   U

6

B   4 m   A

8 m

C

7

P   6 m   U

12 m

D

R

8

E

18 mm

40 mm

F

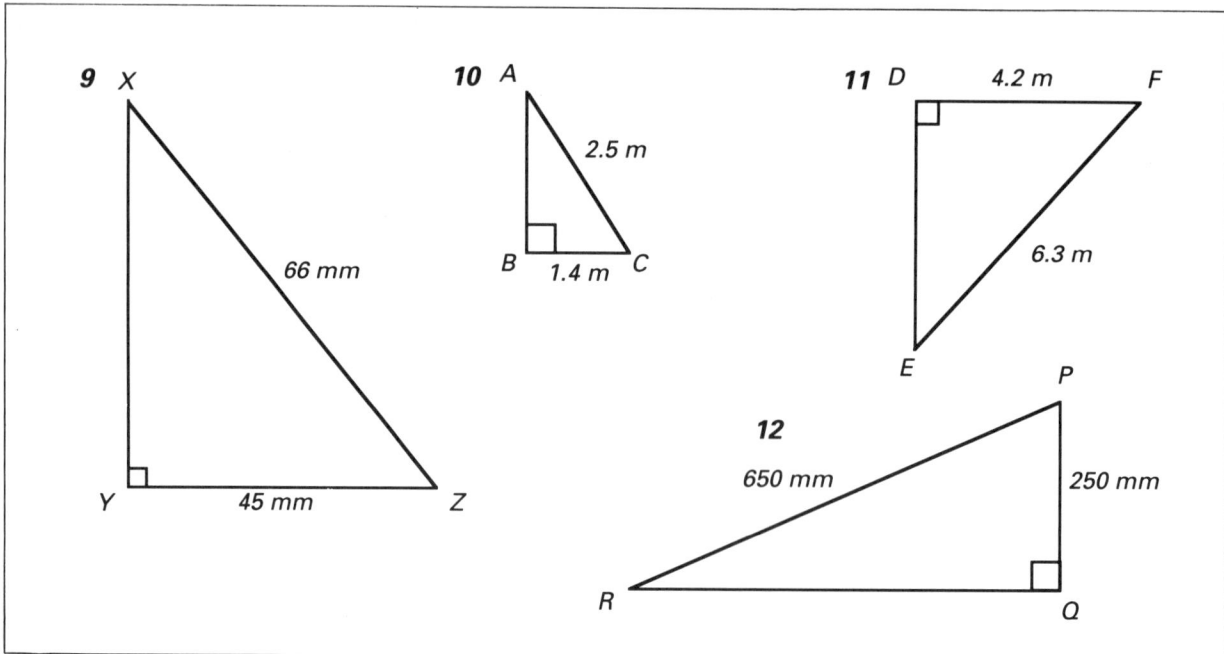

9 X

66 mm

Y   45 mm   Z

10 A

2.5 m

B   1.4 m   C

11 D   4.2 m   F

6.3 m

E

12

650 mm   P

250 mm

R   Q

EXERCISE 5.19

1 A ladder is placed against a wall as shown. How far does it reach up the wall?

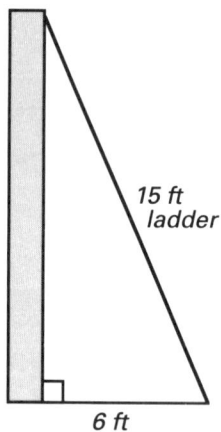

15 ft
ladder

6 ft

2 Find the length of the side AB of the rectangle.

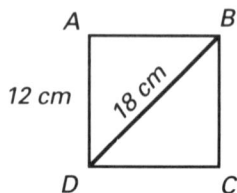

A   B

12 cm   18 cm

D   C

3 The diagonal of a gate is 9 feet long, and the gate is 5 feet high. How wide is it?

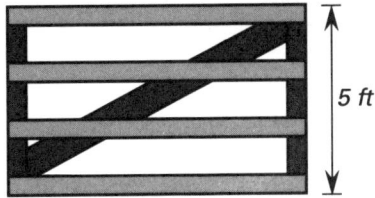

5 ft

4 A kite is tethered on a string 250 metres long, and has travelled a distance of 200 metres horizontally. What is the vertical height of the kite?

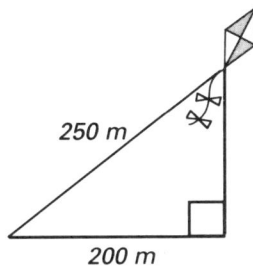

250 m

200 m

Investigation G

This pattern of right-angled triangles is made using the theorem of Pythagoras and what we already know about square roots. Draw the smallest triangle first and add the rest as shown. Continue the pattern for as long as you can. Will the pattern circle around so far that you will eventually end up drawing the triangles on top of each other?

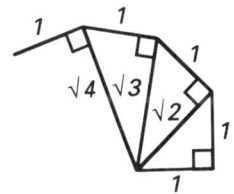

1   1

1

√4 √3

√2

1

1

6. Communication: numbers (2)

The number line

We have met directed numbers several times before, for example, temperatures and number series.

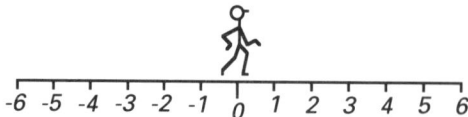

We refer to the series of numbers shown above as the **number line**, and this number line helps us when we are combining directed numbers. Any numbers to the right of 0 are positive numbers, while any numbers to the left of 0 are negative numbers. A negative number needs a minus sign in front of it to show it is negative. As with temperatures, the positive numbers are greater than the negative numbers. The further we move left down the number line, the smaller the numbers are.

Inequalities

$>$ means 'is greater than'
$<$ means 'is less than'

Example 1

Insert an inequality sign ($<$ or $>$) between these numbers: (a) 3   2
(b) -4 4

(a) $3 > 2$ since 3 is greater than 2.
(b) $-4 < 4$ since -4 is less than 4.

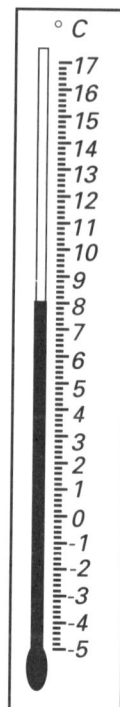

EXERCISE **6.1**

Write down the highest of each pair of temperatures.

1 3°, 7°   **2** −1°, 4°

3 −3°, 0°   **4** −3°, −5°

5 4°, −2°   **6** −5°, −4°

7 8°, −2°   **8** 0°, −5°

9 9°, 3°   **10** −9°, −3°

Insert < or > between each pair of numbers.

11 2 5	**12** 0 −3	**13** −5 −1	**14** 8 2	**15** −1 −3
16 4 −6	**17** −6 −5	**18** −8 4	**19** 2 1	**20** −3 −7
21 5 0	**22** −4 −2	**23** −9 −1	**24** 0 3	**25** −5 −3

EXERCISE **6.2**

Calculate by how much the temperatures have changed.

1 4°, 7°   **2** −2°, 3°   **3** −4°, 0°   **4** −3°, −6°   **5** 5°, −2°

6 −5°, −3°   **7** 7°, −1°   **8** 0°, −4°   **9** 8°, 3°   **10** −5°, −1°

Addition of directed numbers

Directed numbers have a **sign**: either a positive (+) or a negative (−) sign in front of each number to tell us on which side of the 0 on the number line they lie. Numbers written without a sign are always taken to be positive. The sign also gives us a direction to take on the number line. Usually we move in a **positive** direction *up* the line, but a **negative** sign tells us to *change* direction.

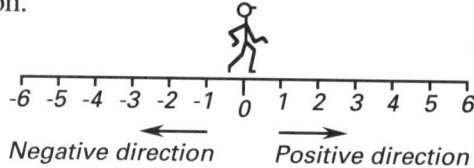

Example 2

$-1 + 3$

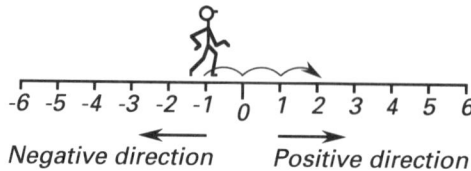

Starting from the first number -1 we move 3 places in the positive direction, arriving at the number 2. So $-1 + 3 = 2$.

Example 3

$2 + (-4)$

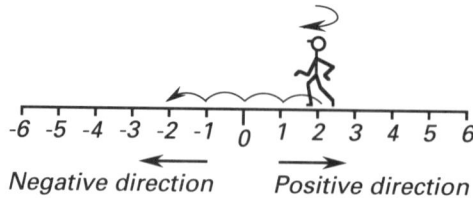

Starting from $+2$ we change direction because of the negative sign, and move in a negative direction 4 places, arriving at the number -2. So $2 + (-4) = -2$.

Example 4

$-1 + (-2)$

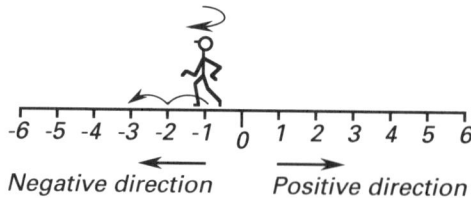

Starting from -1 we change direction because of the negative sign in front of the 2, and move in a negative direction 2 places, arriving at -3. So $-1 + (-2) = -3$.

EXERCISE 6.3

Work out:

1 $3+4$	**2** $-3+4$	**3** $2+(-3)$	**4** $8+(-2)$
5 $-1+(-3)$	**6** $-2+-4$	**7** $-4+5$	**8** $6+(-2)$
9 $4+(-6)$	**10** $-2+3$	**11** $3+-5$	**12** $-2+(-4)$
13 $2+(-1)$	**14** $5+(-6)$	**15** $4+(-8)$	**16** $-2+0$
17 $6+2$	**18** $-5+(-1)$	**19** $1+(-4)$	**20** $-5+(-2)$
21 $5+(-8)$	**22** $-1+3$	**23** $-2+9$	**24** $4+(-1)$
25 $-3+(-7)$			

Subtraction of directed numbers

Example 5

$3-6$

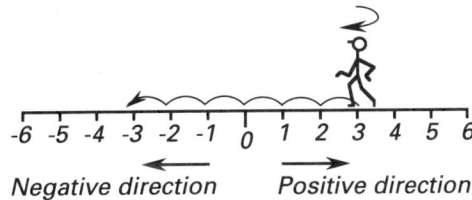

Negative direction   Positive direction

Starting from 3 we change direction because of the negative sign and move in a negative direction 6 places. So $3-6=-3$.

Example 6

$4-(-2)$

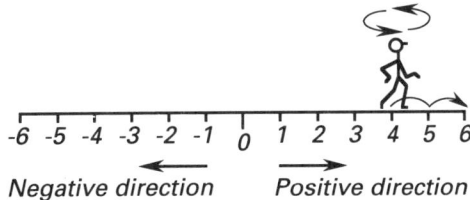

Negative direction   Positive direction

Starting from 4 we change direction because of the first negative sign; we then have to change direction *again* because of a second negative sign, and should now be facing in a positive direction again, moving forward 2 places. So $4-(-2)=6$. The two negative signs together cancel each other out to give a positive sign.

EXERCISE 6.4

Work out:

1 $4-2$	**2** $6-8$	**3** $5-(-1)$	**4** $-2-(-3)$	**5** $9-4$
6 $-8-2$	**7** $-5-(-1)$	**8** $6-2$	**9** $-4-(-5)$	**10** $-2-3$
11 $0-2$	**12** $1-4$	**13** $-5-(-6)$	**14** $3-(-4)$	**15** $7-5$
16 $2-(-9)$	**17** $-4-3$	**18** $2-7$	**19** $-1-(-5)$	**20** $-4-(-3)$
21 $3-2$	**22** $-7-(-4)$	**23** $8-(-6)$	**24** $7-4$	**25** $-5-(-3)$

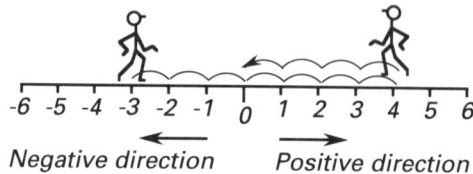

Negative direction   *Positive direction*

Start at -3 facing in a positive direction, move up 7 to 4. Then change direction because of the negative sign, and move 4 back down the line to 0.
So $-3+7-4=0$.

EXERCISE 6.5

Work out:

1 $-5+7-1$	**2** $8-3-1$	**3** $-3-2+7$	**4** $-5+2-3$	**5** $-4+7-9$
6 $4+5-2$	**7** $-8+4-5$	**8** $-6-7+1$	**9** $2-8-3$	**10** $5+1-3$
11 $-4-9+7$	**12** $8+2-3$	**13** $-5-4-2$	**14** $-6-8+4$	**15** $-1+5-6$
16 $9-7+4$	**17** $-3+9-4$	**18** $-7-1-5$	**19** $-2-6-8$	**20** $-5-3+1$
21 $-4+2+9$	**22** $12-2+6$	**23** $8-5-3$	**24** $-6-2-1$	**25** $-10+7+3$

Zeros

Example 8

4 . . . . = 0. What is missing?

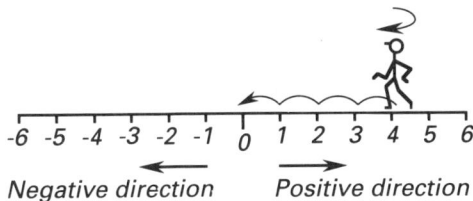

Negative direction   Positive direction

If we start at 4 and want to get back to 0 then we first have to change direction, and then move back 4, so the missing number must be −4. So $4 - 4 = 0$.

EXERCISE 6.6

Find the missing numbers.

1 $-2 + 2 = \ldots\ldots$ **2** $7 \ldots\ldots = 0$ **3** $3 - 3 = \ldots\ldots$

4 $-6 \ldots\ldots = 0$ **5** $3 \ldots\ldots = 0$ **6** $-8 \ldots\ldots = 0$

7 $6 \ldots\ldots = 0$ **8** $1 \ldots\ldots = 0$ **9** $-5 \ldots\ldots = 0$

10 $9 \ldots\ldots = 0$

Multiplication of directed numbers

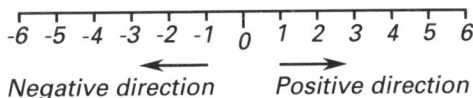

Negative direction   Positive direction

Example 9

3×-4

Multiplying the number parts gives us $3 \times 4 = 12$, but is the answer positive or negative? The first number starts us off facing in a positive direction, but the negative sign tells us we have to change direction, so the answer is negative: −12. So $3 \times -4 = -12$.

Example 10

-3×6

-3×6 is the same as 6×-3, and as this is similar to the previous example, we know the answer will be negative, so $-3 \times 6 = -18$.

Example 11

-2×-4

The first negative sign tells us to change direction and face towards the negative part of the number line. The second negative sign tells us to change direction *again* to face in a positive direction, so the answer will be $+8$. So $-2 \times -4 = 8$.

Investigation A

Try the following combinations of multiplication several times each with directed numbers of your own choice:

positive number $\times$ positive number $=$? number
positive number $\times$ negative number $=$? number
negative number $\times$ positive number $=$? number
negative number $\times$ negative number $=$? number

Can you find a simple rule to use when multiplying directed numbers together?

Investigation B

$(-2)^2 = -2 \times -2 = ?$
$(-2)^3 = -2 \times -2 \times -2 = ?$
$(-2)^4 = -2 \times -2 \times -2 \times -2 = ?$

Work out the answers to these problems, and continue the series up to $(-2)^8$. Compare your answers. Can you see a pattern building up in the numbers and the signs? Can you predict the sign of $(-2)^{11}$?

EXERCISE 6.7

Work out:

1 3×-2	**2** -4×5	**3** -3×-2	**4** 4×-8	**5** 2×9
6 -5×-7	**7** -3×4	**8** 9×-8	**9** -6×-10	**10** 4×-7
11 $(-4)^2$	**12** 12×9	**13** 7×-6	**14** -3×-6	**15** 11×-3
16 -7×-8	**17** $(-5)^2$	**18** 4×-8	**19** $(-4)^2$	**20** 6×5
21 9×-12	**22** -15×3	**23** -10×-9	**24** $(-3)^2$	**25** 13×-3

Division of directed numbers

Exactly the same methods as we have used to multiply directed numbers
can be used in dividing.

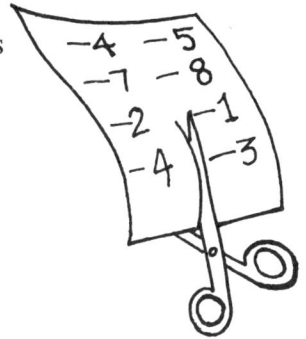

Example 12

Work out $(a)\ 6 \div -2$ $(b)\dfrac{-12}{-3}$.

$(a)\ 6 \div -2 = -3.$

$(b)\dfrac{-12}{-3} = -12 \div -3 = -(-4) = 4.$

Investigation C

Try the following combinations of division several times each with
directed numbers of your own choice:

positive number ÷ positive number = ? number
positive number ÷ negative number = ? number
negative number ÷ positive number = ? number
negative number ÷ negative number = ? number

Can you find a simple rule to use when dividing directed numbers?

EXERCISE 6.8

Work out:

1 $-4 \div 2$ **2** $-8 \div -4$ **3** $-9 \div 3$ **4** $16 \div -4$ **5** $20 \div 5$

6 $-18 \div -6$ **7** $18 \div -9$ **8** $-15 \div 5$ **9** $-36 \div -12$ **10** $-54 \div -9$

11 $64 \div -16$ **12** $-30 \div 15$ **13** $-12 \div -6$ **14** $-66 \div -11$ **15** $28 \div -7$

16 $\dfrac{-84}{12}$ **17** $\dfrac{8}{-2}$ **18** $\dfrac{-56}{-7}$ **19** $\dfrac{-42}{6}$ **20** $\dfrac{-120}{-10}$

EXERCISE **6.9**

Work out:

1 $-3+7$	**2** $-8 \div -4$	**3** $6+7$	**4** $6 \div -2$	**5** $-9-8$
6 $-2 \div 8$	**7** $5 + -3$	**8** $5-8$	**9** 3×-1	**10** $-6 \div -6$
11 $5-1$	**12** -9×7	**13** $4 \div -1$	**14** -10×-1	**15** 12×9
16 $-8 + -1$	**17** $4 - (-3)$	**18** 6×-7	**19** $2-9$	**20** $-5 + -3$
21 $-5+6$	**22** 6×-4	**23** $3+3-10$	**24** $-48 \div 4$	**25** $-3-4+10$

Problem-solving

One of the skills we shall always need to practise is that of solving a problem which involves planning, and a strategy to solve it. We practised some problem-solving skills in Book 3X, and this series of problems will give you further practice. Each problem can be tackled as an investigation.

EXERCISE **6.10**

1 How many blocks are there in stage 1, stage 2, stage 3? Put your results into a table, and extend it to consider stages 4, 5 and 6. From your results so far, can you predict the number of blocks in stages (*a*) 10 (*b*) 15 (*c*) 20?

Stage 1   *Stage 2*   *Stage 3*

2 Repeat the investigation in question 1 for this set of blocks.

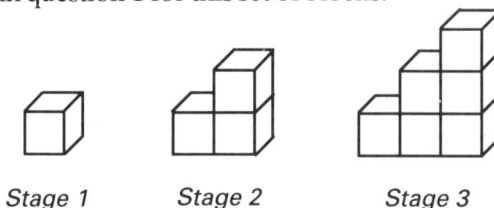

Stage 1   *Stage 2*   *Stage 3*

3 Repeat the investigation in question 1 for this set of blocks.

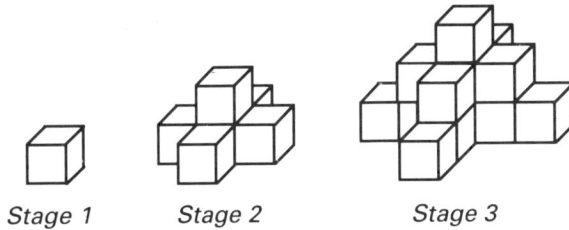

Stage 1   Stage 2   Stage 3

4 A board game is designed such that every 7th square is shaded. How many squares will be shaded on a board measuring (a) 8 × 8 (b) 9 × 9 (c) 10 × 10? (d) Using a table to present your findings, work out the number of squares on a board of size 20 × 20.

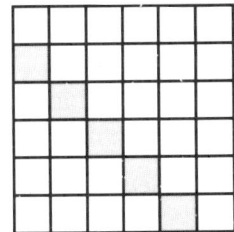

5 The diagram shows the various ways simple dot-matrix printers represent a zero, depending on the height of the character. Find the number of dots in zeros of height 3 to 8 characters. Find an easy way to calculate the number of dots in zeros of height (a) 15 characters (b) 20 characters.

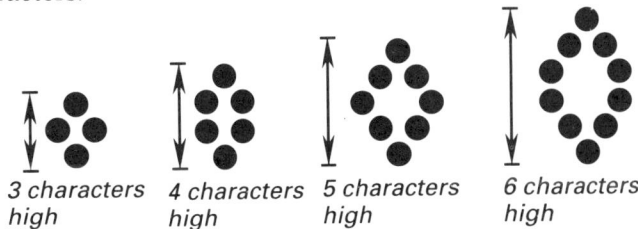

3 characters high   4 characters high   5 characters high   6 characters high

6 Mrs Bailey puts one red sock and one blue sock on a line. How many different ways can these socks be arranged on the line? Try the same problem for 3 and 4 socks of different colours, presenting your solutions in a table. Can you find how many ways there would be for (a) 8 socks (b) 12 socks of different colours?

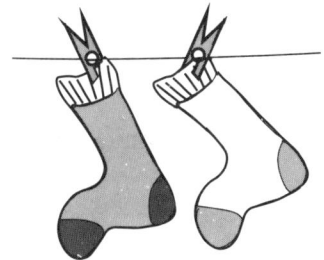

7 A diagram is built up of squares representing steps. What is the perimeter of each diagram? What is the area of each diagram? Find the area and perimeter for diagrams representing 4, 5 and 6 steps. Can you see a pattern developing in the numbers? Predict the area and perimeter for diagrams representing (a) 15 steps (b) 20 steps.

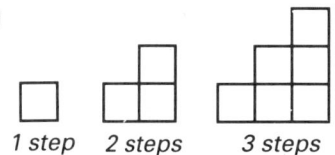

1 step   2 steps   3 steps

8 A man charges £200 for drilling any depth up to the first 2 metres, £250 for any depth up to the next 2 metres, £300 for any depth up to the next 2 metres, and so on. How much will he charge for drilling to depths of (a) 15 m (b) 20 m (c) 25 m?

Investigation D

A pond is surrounded by paving stones, creating a path two stones wide. How many stones will be required for a pond measuring 3×2 stones or 3×4 stones? Try to find a rule to help you work out an answer easily for a pond having any dimensions.

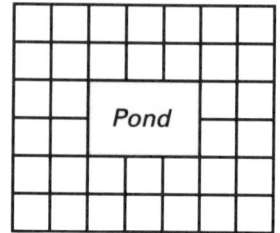

Sets

A group of things which have something in common is called a **set**. Examples could include:

Sets of red books, sharp knives, tin cans.

Sets of people: Kings, American Presidents, mathematics teachers.

Sets of numbers: even numbers, numbers on a dice, house numbers.

We use brackets { } when writing a set.

We could write a set of girl's names as {Ann, Louise, Julie, Lisa}. Each separate part of a set is called an **element**, or a **member**. Ann is an element of the set of girls' names.

Example 13

How could we describe the set {4, 9, 16, 25, 36}?

This is the set of the first five square numbers: $\{2^2, 3^2, 4^2, 5^2, 6^2\}$.

EXERCISE 6.11

Write in brackets the elements of the following sets.

1 {Subjects you are taught}

2 {Even numbers less than 10}

3 {Continents of the world}

4 {Prime numbers less than 20}

5 {Television channels}

6 {The last 8 letters of the alphabet}

7 {Multiples of 6, less than 50}

8 {The colours of your exercise books}

9 {The factors of 36}

10 {The letters and numbers in your postcode}

Describe in your own words these sets.

11 {a, e, i, o, u}

12 {5, 10, 15, 20, 25}

13 {apple, orange, lemon, tomato}

14 {Man, Skye, Skerry, Wight}

15 {Manchester, Liverpool, Newcastle, Birmingham}

16 {1, 8, 27, 64, 125}

17 {35, 36, 37, 38, 39}

18 {red, yellow, blue, green}

19 {Ford, Nissan, Rover, Renault}

20 {11, 13, 15, 17, 19}

100 BEST SETS

Example 14

How many elements has the set {odd numbers less than 20} ?

The odd numbers are 1, 3, 5, 7, 9, 11, 13, 15, 17, 19, so there are 10 odd numbers less than 20, and 10 elements in this set.

EXERCISE 6.12

Write down four elements from each of these sets.

1 {Names of girls}

2 {Animals found in Australia}

3 {Factors of 48}

4 {Pop records}

5 {Months of the year}

6 {Animals found on a farm}

7 {Names of supermarkets}

8 {Teachers in your school}

9 {Names of flowers}

10 {Cities in Great Britain}

How many elements has each of the following sets?

11 {Days of the week}

12 {Players in a football team}

13 {Pupils in your class}

14 {Factors of 20}

15 {Colours of the rainbow}

16 {Planets in the solar system}

17 {Playing cards in a deck}

18 {Cats without tails}

19 {Triangles with four sides}

20 {An athletic relay team}

Comparing sets

Two sets are equal if they have exactly the same members. If $A = \{$blue, black, red, green$\}$ and $B = \{$blue, red, black, green$\}$ then $A = B$. It does not matter if the elements of a set are written in a different order. It is the actual elements of a set which are important.
If $C = \{1, 3, 4, 5, 7, 8\}$ and $D = \{1, 3, 4, 6, 7, 8\}$ then $C = D$.

A set with no elements is called the **empty set** or null set, and is written $\{\,\}$, or given the symbol $\varnothing$.

A shorter way to write down '2 is an element of the set of even numbers' is to write: $2 \in \{$even numbers$\}$.

$\in$ means 'is an element of'
$\notin$ means 'is not an element of'.

For example, knife $\notin \{$things you can drink out of$\}$.

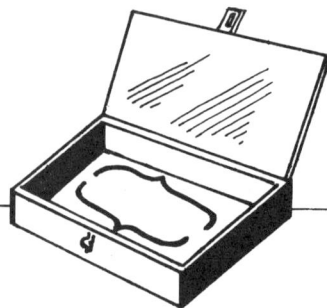

EXERCISE **6.13**

Find out which of the following sets are empty.

1 $\{$People who have swum the Atlantic$\}$ **2** $\{$Names of dogs without tails$\}$

3 $\{$Men who have walked on the moon$\}$ **4** $\{$People over 100 years old$\}$

5 $\{$Pupils more than 2 metres tall$\}$ **6** $\{$Men who can run at more than 16 m.p.h.$\}$

State whether each pair of sets is equal or not equal.

7 $A = \{1, 3, 7, 9, 13, 15\}$; $B = \{1, 3, 7, 8, 13, 15\}$

8 $A = \{$Aries, Taurus, Pisces, Gemini, Virgo$\}$; $B = \{$Pisces, Taurus, Virgo, Aries, Gemini$\}$

9 $A = \{b, f, i, t, g, s, x\}$; $B = \{f, i, v, b, x, s, g\}$

10 $A = \{36, 72, 49, 87, 59, 30\}$; $B = \{49, 30, 87, 36, 59, 73\}$

11 $A = \{$B349FTU, JBU542V, G49XTU$\}$; $B = \{$JBU542V, G49XTU, B349FTU$\}$

12 $A = \{$BSc, DASE, Dip, MA, DPhil$\}$; $B = \{$DPhil, MA, BSc, Dip, DASE$\}$

Write down whether each statement is TRUE or FALSE.

13 Tomato $\in$ {Vegetables}

14 Tea pot $\notin$ {Cutlery}

15 Peas $\notin$ {Fruit}

16 Apples $\in$ {Trees}

17 Tin $\notin$ {Rare metals}

18 Orchids $\in$ {Common flowers}

19 £5 note $\notin$ {Coins}

20 Cup {Crockery from which to drink}

21 Picture $\notin$ {Things you hang on the wall}

22 Video cassette $\notin$ {Items you use with a hi-fi}

Venn diagrams

Sometimes it is clearer to show the elements of a set in a diagram rather than written out in a list. We use Venn diagrams to do this.
If $A =$ {Numbers less than 10} $= \{1, 2, 3, 4, 5, 6, 7, 8, 9\}$, then we can show set A in a Venn diagram:

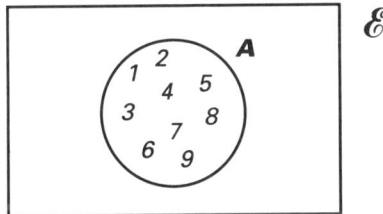

The circle represents the set A itself, and the box around it all the other numbers we may want to consider.

Example 15

From all numbers less than 10, set $B =$ {even numbers}. Show this in a Venn diagram.

$B = \{2, 4, 6, 8\}$.

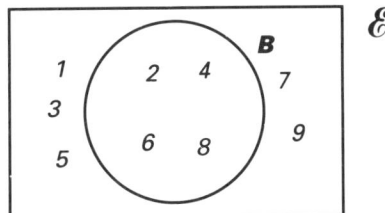

The elements of the set B go inside the circle representing B, and all other numbers we don't need go outside the circle, but written within the rectangle. The rectangle and all the numbers inside it is called the **universal set**, and is given the symbol $\mathscr{E}$.

Hence, from the diagram, the universal set in this case is $\{1, 2, 3, 4, 5, 6, 7, 8, 9\}$.

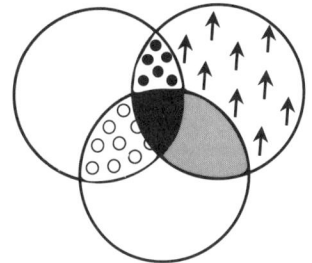

EXERCISE 6.14

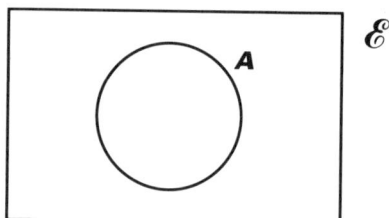

For each question, copy the Venn diagram above and insert the numbers.

1 Universal set = {5, 10, 15, 20, 25, 30, 35, 40}; A = {5, 15, 25, 35}

2 Universal set = {Even numbers less than 20}; A = {2, 4, 6, 8}

3 Universal set = {First ten letters of the alphabet}; A = {a, e, i}

4 Universal set = {Numbers between 40 and 50}; A = {Odd numbers between 40 and 50}

5 Universal set = {Factors of 48}; A = {Factors of 16}

Combining sets

There are two common ways in which we combine any two sets.

First, we may be interested in those elements which are in both the sets.

$$A = \{1, 3, 4, 7, 8\} \qquad B = \{6, 7, 8, 9\}$$

The set of numbers in both sets is {7, 8}. This is called the **intersection** of sets A and B, and we can show the intersection on a Venn diagram:

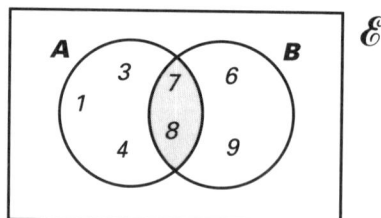

This can be represented using the notation $A \cap B$, and we write $A \cap B = \{7, 8\}$. In the diagram the two sets A and B are shown overlapping, or intersecting, with the numbers 7 and 8, the two elements in common, in the middle. Hence the shaded area represents the intersection of the two sets.

Second, we may be interested in all the elements of the two sets. The set of all the elements of sets A and B is $\{1, 3, 4, 6, 7, 8, 9\}$. Notice we only write the numbers 7 and 8 once in this set. This set of all the elements is called the **union** of the sets A and B, and we can also show it on a Venn diagram:

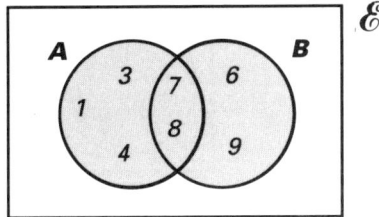

This can be represented using the notation $A \cup B$, and we write

$$A \cup B = \{1, 3, 4, 6, 7, 8, 9\}$$

The shaded area represents the union of the two sets.

EXERCISE **6.15**

1 Write down the elements of (a) set A (b) set B (c) those elements in both sets: $A \cap B$ (d) all the elements in the sets: $A \cup B$.

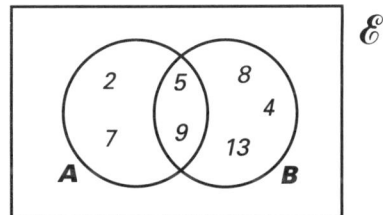

2 Write down the elements of (a) set P (b) set Q (c) those elements in both sets: $P \cap Q$ (d) all the elements in the sets: $P \cup Q$.

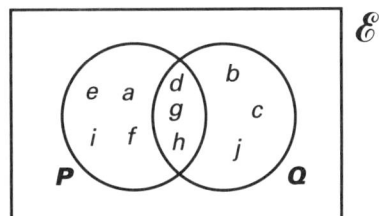

3 Write down the elements of (a) set X (b) set Y (c) those elements in both sets: $X \cap Y$ (d) all the elements in the sets: $X \cup Y$.

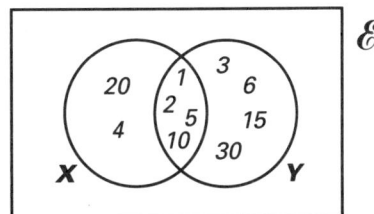

4 Write down (*a*) all those who study history (*b*) all those who study geography (*c*) all those who study both history and geography (*d*) all those who study humanities (either history or geography).

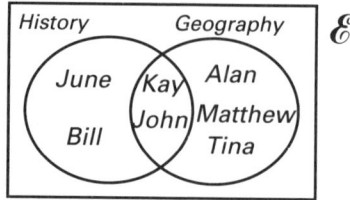

5 Write down (*a*) the multiples of 3 (*b*) the multiples of 4 (*c*) all those numbers which are multiples of both 3 and 4 (*d*) all those numbers which are multiples of either 3 or 4.

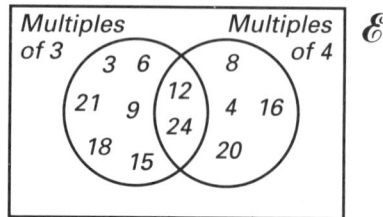

Example 16

$A = \{30, 32, 45, 48, 50\}$; $B = \{35, 40, 45\}$.

Draw a Venn diagram to help you find (*a*) the set containing all the elements of the two sets, (*b*) the set with only those elements which are in both the sets.

First of all find those elements which are in both the sets: $\{45\}$, and add this to the middle of the diagram:

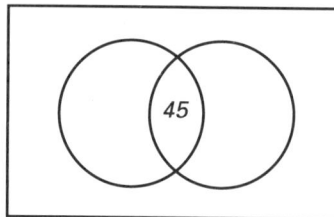

We can then add the other elements of set A, and the other elements of set B, to the diagram:

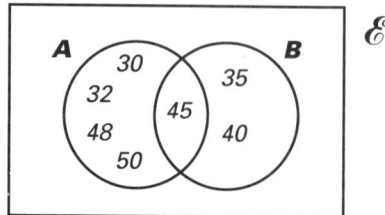

From the diagram we can see that

(*a*) the union of A and B is the set $\{30, 32, 35, 40, 45, 48, 50\}$

(*b*) the intersection of A and B is the set $\{45\}$.

EXERCISE **6.16**

Draw Venn diagrams for each pair of sets. For every question write down
(*a*) the elements in both sets (*b*) all the elements of each pair of sets.

1 $A = \{1, 4, 6, 7, 9, 11, 15\}$; $B = \{3, 6, 9, 12, 15\}$

2 $A = \{c, h, d\}$; $B = \{e, c, s, t, r\}$

3 $A = \{$Numbers less than 12$\}$; $B = \{$Odd numbers between 6 and 18$\}$

4 $A = \{3, 4, 5, 6, 9, 12, 13\}$; $B = \{6, 7, 8, 9, 10, 11, 12\}$

5 $A = \{31, 33, 35, 37, 39\}$; $B = \{32, 37, 38, 40\}$

6 Pupils travelling to school by train = {Bob, Jenny, Vivek, Alan, Trish, Ann};
pupils walking to school = {Remi, Alan, Sadiah, Tina, Kathy, Mike, Ann}.

Example 17

Universal set = {All numbers less than 20}

$A = \{2, 5, 7, 9, 12, 15\}$ $B = \{4, 9, 10, 12, 15, 18, 19\}$

Draw a Venn diagram to represent these sets.

Begin by drawing the diagram with just those elements which are in both
A and B in the middle:

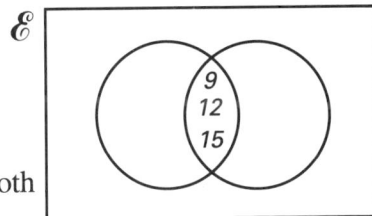

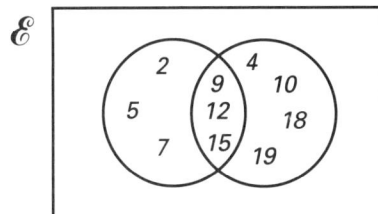

Then add the other elements of the sets A and B:

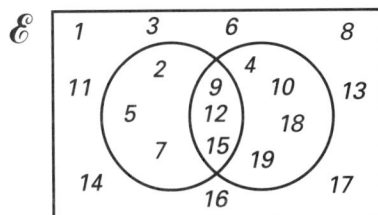

Finally add all the other numbers in the universal set which have not
been included in the diagram already:

EXERCISE 6.17

Draw Venn diagrams to represent the sets in each question.

1 Universal set = {All whole numbers less than 12},
$A = \{1, 2, 4, 6\}$, $B = \{4, 5, 6, 7\}$.

2 Universal set = {The first 16 letters of the alphabet},
$A = \{a, b, d, e, f, g\}$, $B = \{b, e, f, i, k\}$.

3 Universal set = {All numbers less than 20},
$A = \{\text{even numbers}\}$, $B = \{\text{numbers less than 10}\}$.

4 Universal set = {Numbers between 30 and 40},
$A = \{31, 33, 35\}$, $B = \{32, 37, 38\}$.

5 Universal set = $\{a, e, f, g, h, m, p, q, r, x, y\}$,
$A = \{e, f, h, q, x\}$, $B = \{g, h, m, x, y\}$.

6 Universal set = {Numbers between 60 and 70},
$A = \{\text{odd numbers}\}$, $B = \{61, 63, 65, 67\}$.

Investigation E

Universal set = {All pupils in your class}.

$A = \{\text{All the girls}\}$ $B = \{\text{All pupils with fair hair}\}$

Draw a Venn diagram to represent this situation.
    Suggest and draw other Venn diagrams to show the differences between the members of your class.

Example 18

The Venn diagram represents the results of a survey to find how many pupils come to school by train and by walking. The numbers in the Venn diagram show the number of elements in each set.
(*a*) How many come by train? (*b*) How many walk to school? (*c*) How many children walk to school and also come by train? (*d*) How many children are there in the class? (*e*) How many children do not walk to school?

(*a*) $3 + 8 = 11$. (*b*) $8 + 15 = 23$. (*c*) 8. (*d*) $3 + 8 + 15 + 4 = 30$.
(*e*) $30 - 8 - 15 = 7$.

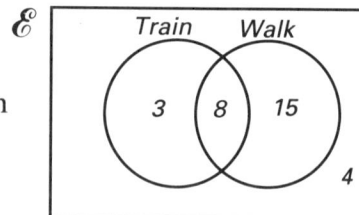

EXERCISE **6.18**

Each Venn diagram shows the number of elements in each set.

1 The diagram shows the number of *TV Times* and *Radio Times* being delivered to houses in a street.

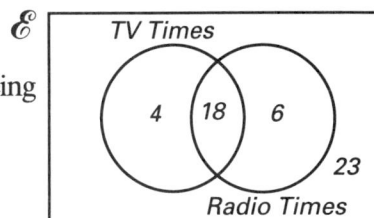

(*a*) How many have both *TV Times* and *Radio Times* delivered?

(*b*) How many have neither magazine delivered?

(*c*) How many houses do not have the *TV Times* delivered?

(*d*) Find the number of houses in this street.

$\mathcal{E}$

TV Times

4 18 6

23

Radio Times

2 The diagram shows the number of houses in a road requiring milk or dairy produce from the milkman.

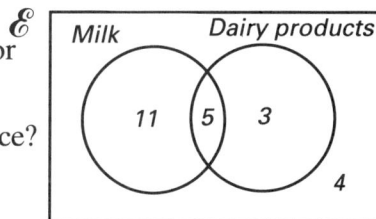

(*a*) How many houses have deliveries of both milk and dairy produce?

(*b*) How many houses have only milk delivered to them?

(*c*) Find the number of houses in the road.

(*d*) At how many houses does the milkman not call?

$\mathcal{E}$

Milk   Dairy products

11 5 3

4

3 The Venn diagram shows the results of a survey to compare the number of children preferring cheese and onion crisps, or salt and vinegar crisps.

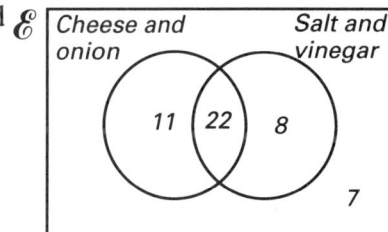

(*a*) Which flavour of crisp was most popular?

(*b*) How many children were asked?

(*c*) How many children preferred a different flavour of crisp?

$\mathcal{E}$

Cheese and   Salt and
onion   vinegar

11 22 8

7

4 One hundred people were questioned to find out their choice of main holiday last year.

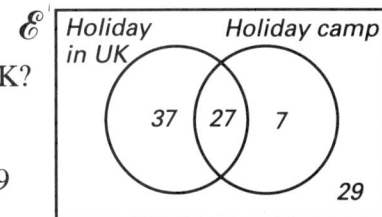

(*a*) How many people took their holiday in the UK?

(*b*) How many took their holiday in a holiday camp outside the UK?

(*c*) Find the number of people who took their holiday abroad.

(*d*) What type of holiday might have been taken by the group of 29 people in the diagram?

$\mathcal{E}$

Holiday   Holiday camp
in UK

37 27 7

29

EXERCISE 6.19

Draw a Venn diagram to help you answer each question.

1 Thirty-six householders were asked whether they had a dog or a cat at home. Three had just a dog, four had both a dog and a cat, and five just a cat.

(a) How many householders had dogs?

(b) How many householders had cats?

(c) Find the number of people asked who had neither dogs nor cats.

(d) Which were most popular: dogs or cats?

2 In a school year group of 150 pupils, there were German and Spanish language classes for those wanting to study a second foreign language. There were 107 pupils who did not study a second foreign language. A total of 35 studied Spanish, and 18 were in the German class.

(a) How many took both Spanish and German?

(b) How many took only Spanish?

(c) How many took only German?

(d) How many children in the year group did not take Spanish?

3 There were a number of vacancies for apprentices in a company requiring craft apprentices, technical apprentices and other types of apprentice. The company received 620 applications for its jobs. Of these 423 were for jobs other than the craft and technical apprentices, 84 applications were for the craft vacancies only, and 63 were for the technical vacancies only.

(a) How many applied for both the craft and technical vacancies?

(b) How many applications in total did the company receive for the craft posts?

(c) How many applications in total did the company receive for the technical jobs?

(d) Find the number of people who were not interested in the craft vacancies.

4 One hundred pupils were questioned to find their preference with regard to tea or coffee. Their answers showed that 5 preferred tea alone, 28 liked both tea and coffee. There were 56 pupils who did not like tea or coffee.

(*a*) How many pupils liked just coffee?

(*b*) What was the total number of pupils who would drink tea?

(*c*) What was the total number of children who would drink either tea, coffee, or both?

5 During thirty days of winter a record was kept to find whether it rained or snowed. On seven days it just rained, on four days it just snowed, and on a further five days it rained and snowed on the next day.

(*a*) Find how many of the thirty days were dry.

(*b*) On how many days did it snow?

(*c*) On how many days did it rain?

(*d*) Find on how many days it was wet with either rain or snow.

6 A sample of fifty library books were examined to find whether they contained colour or black and white photographs. Twelve had just black and white, six had colour only, and twenty-four had no photographs at all.

(*a*) How many books contained both black and white and colour photographs?

(*b*) In how many books could colour photographs be found?

(*c*) In how many books could black and white photographs be found?

(*d*) How many of the books contained photographs?

Investigation F

This design is to be found on many Persian bowls.
These patterns can be divided into two sets: those with an even number of points, and those with an odd number of points. Draw several other designs like this one, and write down as many details as you can find about each set of patterns. You might like to consider these questions for each set.

(*a*) How many cross-over points has the design?

(*b*) Can it be drawn as a continuous line?

(*c*) How many separate spaces are there contained within the lines of the design?

7. Communication: graphics (1)

Coordinates

In previous work in mathematics, and in many other subjects, you have met the idea of plotting points on a graph. The points may have represented the heights and weights of a class of pupils, or the temperature of a patient taken every three hours. In order to extend graphical work further, let us remind ourselves of the basic ideas, using x and y axes.

In the diagram, the **x coordinate** of A is 3, and the **y coordinate** of A is 2. We can simplify this by stating:

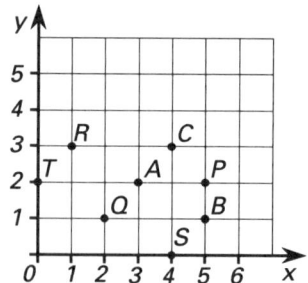

A has coordinates (3,2), or, even shorter, A is (3,2).

Similarly, B is (5,1) and C is (4,3).

Example 1

Write down the coordinates of P, Q, R, S and T in the diagram.

P is (5,2), Q is (2,1) and R is (1,3).
S lies on the x axis, so its y coordinate is zero. Hence S is (4,0).
Similarly, as T is on the y axis, its x coordinate is zero; so T is (0,2).

It is very important to make sure that you write down the x coordinate *first*, followed by the y coordinate, with a 'comma' between them, and brackets round both numbers.

EXERCISE 7.1

1 (*a*) Copy this grid on squared paper.
(Remember to write x at the end of the x axis, and y at the end of the y axis.)

(*b*) What are the coordinates of the points J,K and L?

(*c*) The points J, K and L are three of the corners of a rectangle. Mark the fourth corner, and label it M.
What are the coordinates of M?

(*d*) Draw in the two diagonals JK and LM. They meet at N.

(*e*) What are the coordinates of N?

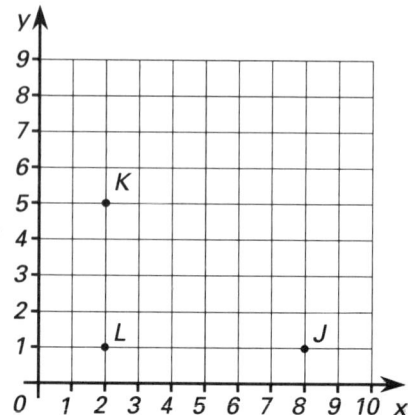

(*f*) Draw a straight line joining the point (8,0) to the point (0,8).

(*g*) Through which of the five lettered points does the line pass?

(*h*) Write down the coordinates of any two other points on this line.

2 (*a*) Draw another grid, just like the one in question 1.

(*b*) P is (0,2). What is Q?

(*c*) R is the point (8,6).
Mark R on your diagram.

(*d*) S is the fourth corner of the square PQRS. What are its coordinates?

(*e*) The centre of the square is the point T. Find its coordinates.

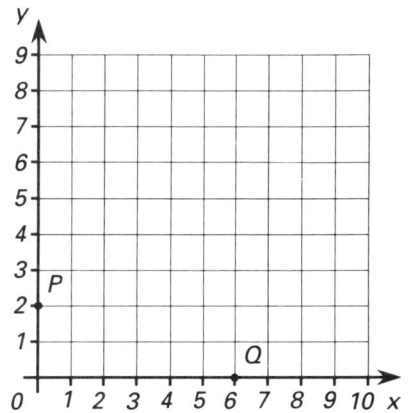

Investigation

On a grid with an *x* axis going up to about 16, and a *y* axis going up to about 8, write your initials, using straight line segments.

Now write down the coordinates of those points in the letters, in the order in which they must be joined so as to spell out your initials. Some points may have to be drawn through twice.

You can extend this to your first name spelt in full. (It's easy if you are called Ann or Ian, harder if you are called Christine or Christopher!)

Negative coordinates

If we extend the *x* axis backwards beyond the origin (0,0), we can label it with negative numbers; similarly, we can extend the *y* axis below the origin. This enables us to work with negative coordinates.

Example 2

What are the coordinates of the points U, V, W, X, Y and Z in the diagram?

U is $(-2,3)$, V is $(-4,2)$, W is $(4,-1)$, X is $(0,-3)$, Y is $(-3,-4)$ and Z is $(-4,0)$.

Remember that the *x* coordinate measures the position *across* and the *y* coordinate measures the position *up* (or *down*).

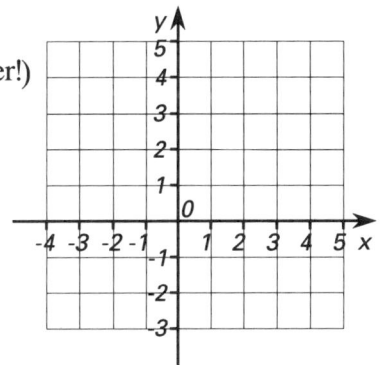

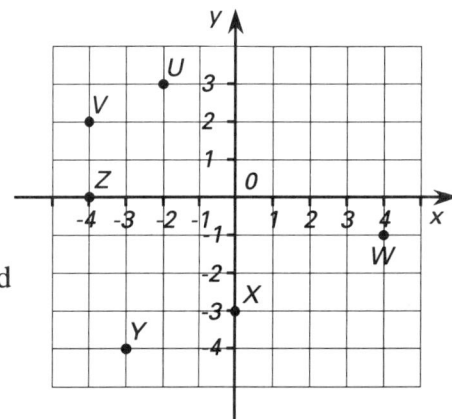

EXERCISE 7.2

1 (a) Draw a grid, with the x axis going from −6 to +6, and the y axis going from −3 to +5. Mark the points A, B, C and D on your diagram.

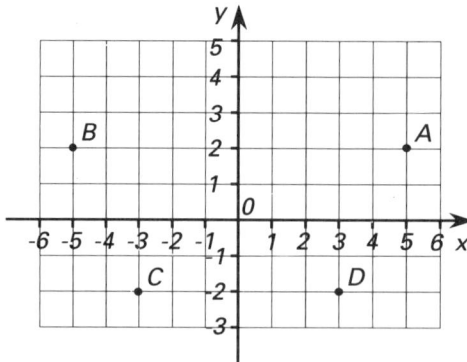

(b) What are the coordinates of A, B, C and D?

(c) At what point does the line AD meet the x axis? Draw in the line.

(d) At what point does the line BC meet the x axis? Draw in the line.

(e) At what point does the line CD meet the y axis? Draw in the line.

(f) E is the point (2,4). Mark E on your diagram.

(g) If the diagram is folded along the y axis, A would land on B. Similarly C would land on D. What are the coordinates of the point on which E would land?

(h) Write down the coordinates of any point that would land on itself if the diagram is folded along the y axis.

2 (a) The point R has coordinates (−2,3). What are the coordinates of S and T, the other two corners of the triangle?

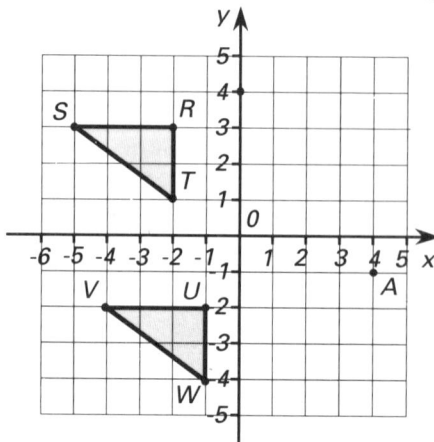

(b) The triangle is slid across and down, without turning, to UVW. What are the coordinates of U, V and W?

(c) UVW is slid across and up, without turning, to ABC, so that U finishes up at A, which has coordinates (4,−1). Draw the position of the triangle, and write down the coordinates of B and C.

(d) Triangle RST is slid, without turning, so that S lands on the point (0,4). What are the new coordinates of the other two corners of the triangle?

3 (a) Draw a grid with axes from −4 to +5.

(b) A is (2,5), B is (0,1), C is (4,1) and D is (1,−1). Plot these points on your grid.

(c) Join A to B with a straight line, and continue the line past B. Join C to D, similarly, so that the two lines meet.

(d) What are the coordinates of the point where these two lines meet?

Straight-line graphs

In Chapter 4 we looked at examples of a rule (formula) to work out some values. Following on from that idea, we can draw a graph which represents a rule. From such a graph, we can read off values without having to calculate them each time.

Example 3

Draw the graph of the equation $y = 2x + 3$, taking values of x from 0 to 5.

If we begin by drawing the x axis from 0 to 5, we find that we do not know how high to draw the y axis. A good idea would be to work out the range of y values first, so that we know the range of the y axis.

When $x = 0$, $y = (2 \times 0) + 3 = 0 + 3 = 3$.
So one point on the graph will be the point with coordinates (0,3).
When $x = 1$, $y = (2 \times 1) + 3 = 2 + 3 = 5$, i.e. (1,5) is on the graph.
When $x = 2$, $y = (2 \times 2) + 3 = 4 + 3 = 7$, i.e. (2,7) is on the graph.
When $x = 3$, $y = (2 \times 3) + 3 = 6 + 3 = 9$, i.e. (3,9) is on the graph.
When $x = 4$, $y = (2 \times 4) + 3 = 8 + 3 = 11$, i.e. (4,11) is on the graph.
When $x = 5$, $y = (2 \times 5) + 3 = 10 + 3 = 13$, i.e. (5,13) is on the graph.

We can now see that the y axis needs to go from 3 up to 13. (We may as well take it from 0.) So we draw the axes, and plot the points that we have worked out.

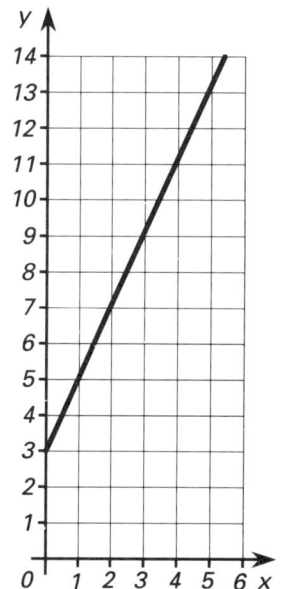

Using a table

It is usually more straightforward, and causes less errors, if we work out the values using a table.

First write the x values in a row (let us use the equation above as an example).

x 0 1 2 3 4 5

Now work out, for each value of x, the corresponding value of $2x + 3$. This can be done in two parts: first, work out $2x$ for each value of x.

x	0	1	2	3	4	5
$2x$	0	2	4	6	8	10

Then add 3 to each result. Set out the table like this:

x	0	1	2	3	4	5
$2x$	0	2	4	6	8	10
$+3$	3	3	3	3	3	3
y	3	5	7	9	11	13

From this table we can see some patterns, which can help to check our work. This example is quite simple, but with more complicated equations the 'table' method can be very useful.

Example 4

Make up a table of values, and then draw the graph of the equation $y = 3x + 4$, for values of x from $x = 0$ to $x = 8$.

The table will look like this:

x	0	1	2	3	4	5	6	7	8
$3x$	0	3	6	9	12	15	18	21	24
$+4$	4	4	4	4	4	4	4	4	4
y	4	7	10	13	16	19	22	25	28

The y axis on the graph will need to go up to 28.

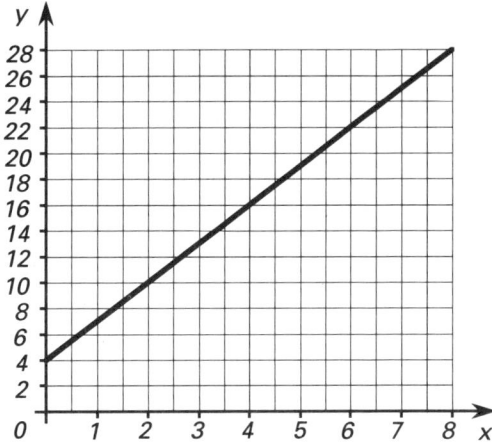

(The scales on each axis are different but this does not affect the reading of values or the plotting of points; it distorts the overall shape of the line by squashing or stretching it in one direction or the other.)

Example 5

Draw the line $y = x - 2$, taking x values from 0 to 7.

In this case, where there is just one x, we need not insert an extra line; we just need a row of -2's, thus:

x	0	1	2	3	4	5	6	7
-2	-2	-2	-2	-2	-2	-2	-2	-2
$y = x - 2$	-2	-1	0	1	2	3	4	5

Here, the y axis needs to go from -2 to $+5$.

We have in effect *added* (-2) to each x value. It will reduce errors in more complicated examples if you stick to the principle of *adding* the figures in the table, in this example, by thinking of the equation as $x + (-2)$.)

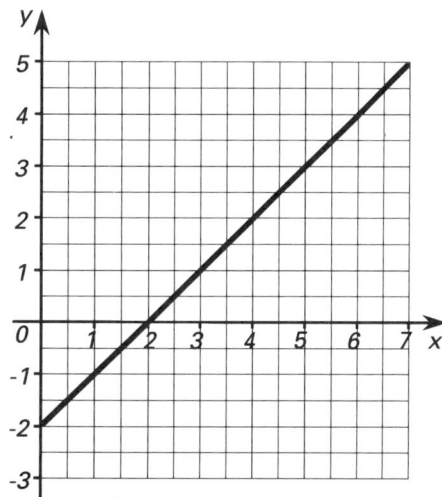

EXERCISE **7.3**

Make up a table for each of these equations. In each case, take values of x from 0 to 6. Do not draw the graphs, just work out the y values in the table, corresponding to the x values of the points that you would need to plot.

1 $y = x + 1$ **2** $y = x + 6$

3 $y = 2x - 1$ **4** $y = 10x + 9$

5 $y = 4x - 7$ **6** $y = \dfrac{x}{2} + 5$

7 $y = \dfrac{x}{4} - 1$ **8** $y = x - 4$

9 $y = 0.1x + 12$ **10** $y = 6x - 9$

The next four examples show other variations in the way equations of straight lines can be written.

Example 6

Draw the line represented by the equation $y = \dfrac{x}{2} + 3$. Take x values from 0 to 10.

There are clearly going to be fractions (or decimals) here. It is usually more convenient in a table to use decimals rather than fractions, rounding them where necessary.

x	0	1	2	3	4	5	6	7	8	9	10
$\dfrac{x}{2}$	0	0.5	1	1.5	2	2.5	3	3.5	4	4.5	5
$+3$	3	3	3	3	3	3	3	3	3	3	3
$y = \dfrac{x}{2} + 3$	3	3.5	4	4.5	5	5.5	6	6.5	7	7.5	8

The y axis goes from 3 to 8 (we may as well start from 0).

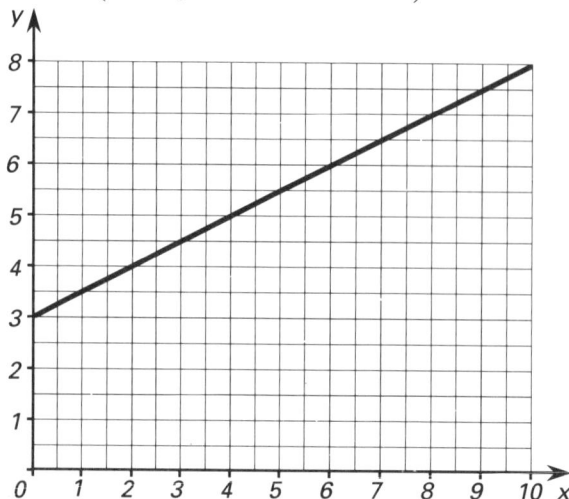

Example 7

Draw the line $y = 2(x + 1)$, for values of x from 0 to 8.

The table will be:

x	0	1	2	3	4	5	6	7	8
$x + 1$	1	2	3	4	5	6	7	8	9
$y = 2(x + 1)$	2	4	6	8	10	12	14	16	18

Here the y axis needs to go from 2 to 18.

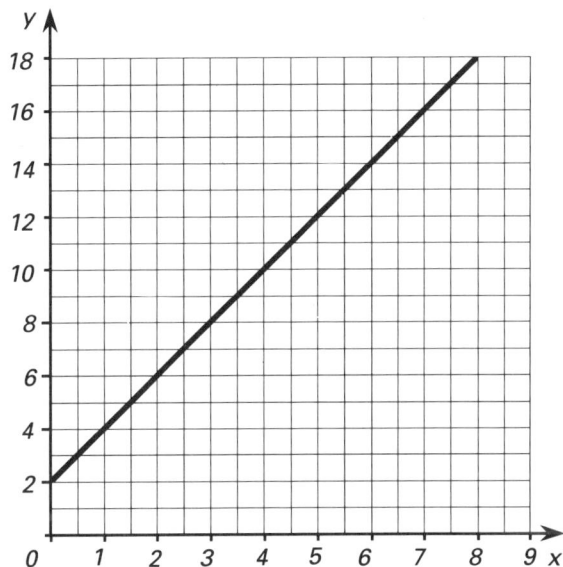

Example 8

Taking values of x from $x = 0$ to $x = 12$, draw the graph of the equation $y = 10 - x$.

As in Example 3, we can look on this as $y = 10 + (-x)$.

x	0	1	2	3	4	5	6	7	8	9	10	11	12
10 $-x$	10 0	10 -1	10 -2	10 -3	10 -4	10 -5	10 -6	10 -7	10 -8	10 -9	10 -10	10 -11	10 -12
y	10	9	8	7	6	5	4	3	2	1	0	-1	-2

The y axis needs to go from -2 to 10.

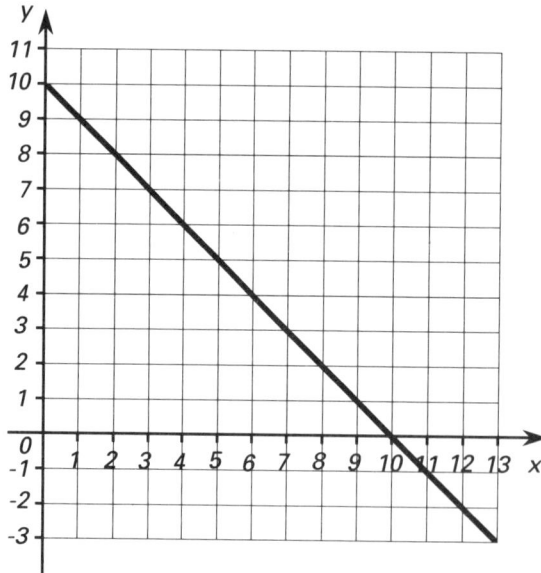

Example 9

Draw the graph of the equation $y = 15 - 2x$, for x values ranging from 0 to 8.

As before, the equation can be written as $y = 15 + (-2x)$.

x	0	1	2	3	4	5	6	7	8
15	15	15	15	15	15	15	15	15	15
$-2x$	0	-2	-4	-6	-8	-10	-12	-14	-16
$y = 15 - 2x$	15	13	11	9	7	5	3	1	-1

This time the y axis goes from -1 to 15.

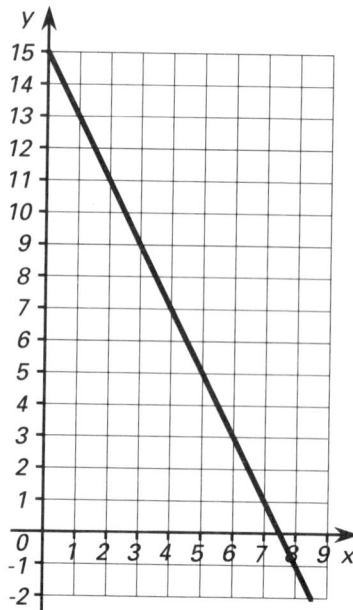

The only other extension that we need to consider at this stage is to have some *negative x* values; from -3 to $+5$, for example.

Example 10

Draw the graph of $y = 3x + 5$, for values of x going from -3 to $+4$.

We begin by writing the x row:

$$x \quad -3 \quad -2 \quad -1 \quad 0 \quad 1 \quad 2 \quad 3 \quad 4$$

Next, the $3x$ row: but when you enter the numbers start from the other end (the '4' end). That is, start by entering 12 under the 4, then 9 under the 3, and so on, until you reach the 0:

x	-3	-2	-1	0	1	2	3	4
$3x$				0	3	6	9	12

You can now see the pattern: from 12 to 9 to 6, etc., is 3 *less* each time, so the next entry will be -3, followed by -6 and -9.

x	-3	-2	-1	0	1	2	3	4
$3x$	-9	-6	-3	0	3	6	9	12

This pattern helps to confirm that $3 \times (-2) = -6$, etc.
Now completing the table will give:

x	-3	-2	-1	0	1	2	3	4
$3x$	-9	-6	-3	0	3	6	9	12
5	5	5	5	5	5	5	5	5
$y = 3x + 5$	-4	-1	2	5	8	11	14	17

The y axis thus goes from -4 to 17. (Note the different scales.)

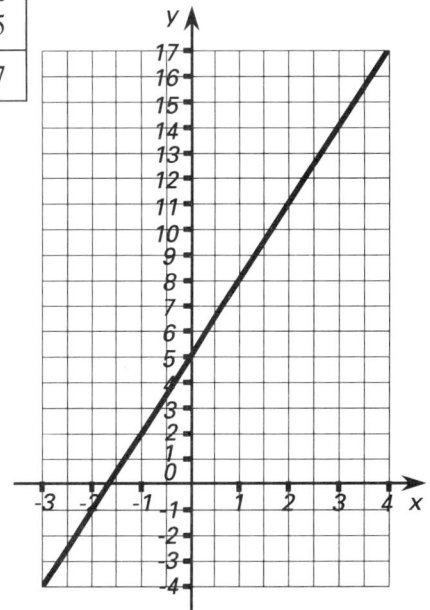

EXERCISE 7.4

For values of x from -3 to $+5$, make up a table and hence draw the graph of each of the equations.

1 $y = x + 2$ **2** $y = 3x - 1$

3 $y = 3(x - 1)$ **4** $y = \dfrac{x + 5}{2}$

5 $y = 2x - 5$ **6** $y = 8 - x$

7 $y = 16 - 3x$ **8** $y = \dfrac{x + 30}{10}$

9 $y = 1.8x + 32$ **10** $y = \dfrac{x}{2} + \dfrac{1}{2}$

Non-linear equations

Draw the graph of $y = \dfrac{24}{x}$, for values of x from 1 to 6.

We have to divide 24 by each x value in order to give us the y value. The table will look like this:

x	1	2	3	4	5	6
$y = \dfrac{24}{x}$	24	12	8	6	4.8	4

The points to plot are $(1,24), (2,12), (3,8), (4,6), (5,4.8)$ and $(6,4)$.

The graph is not a straight line, but a smooth curve. The points are *not* joined by a series of straight lines, but by as smooth a curve as possible. (One way of checking for smoothness is to hold your book at eye level, and follow the curve round by gradually turning your book.)

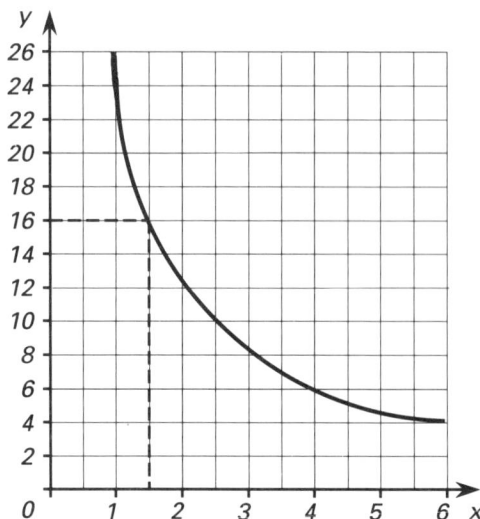

There is quite a gap between the first two points. You can always work out another point, if you wish, to help you to draw the curve more accurately. Let us take $x = 1.5$; the y value will be $\frac{24}{1.5} = 16$. So you can plot the point $(1.5, 16)$, and hence draw a better curve.

Example 11

Make up a table and draw the graph of the equation $y = \frac{36}{x+2}$, taking values of x from 0 to 6. Try to draw as smooth a curve as you can.

The table will be:

x	0	1	2	3	4	5	6
$x+2$	2	3	4	5	6	7	8
$y = \frac{36}{x+2}$	18	12	9	7.2	6	5.1	4.5

and the graph will look like this.

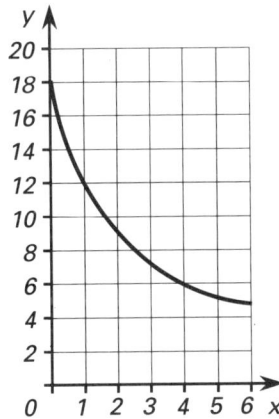

EXERCISE 7.5

Make up a table, and draw the graphs of the following equations. The range of values of x is given for each question.

1 $y = \frac{12}{x}$: x values from 1 to 8.

2 $y = \frac{60}{x+3}$: x values from −1 to 5.

3 $y = \frac{10}{x} + 4$: x values from 1 to 8.

4 $y = \frac{30}{x}$: x values from 1 to 6.

5 $y = \frac{24}{x} - 4$: x values from 1 to 6.

Example 12

Draw the graph of $y = x^2$ for values of x ranging from
(a) $x = 0$ to $x = 5$, (b) $x = -5$ to $x = +5$.

(a) As usual, make up a table.

x	0	1	2	3	4	5
$y = x^2$	0	1	4	9	16	25

Now draw the graph, with y values from 0 to 25.

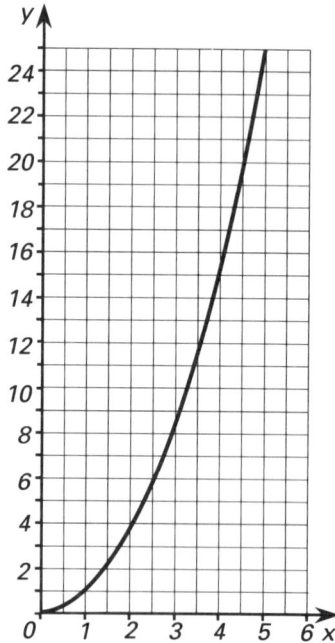

(b) When we have x^2 in the equation, we must remember that if the x value is *negative*, it will become *positive* when it is squared. For example, $(-4)^2 = (-4) \times (-4) = +16$. In fact the square of *any* number is *positive*.
   Make up the table:

x	-5	-4	-3	-2	-1	0	1	2	3	4	5
$y = x^2$	$+25$	$+16$	$+9$	$+4$	$+1$	0	$+1$	$+4$	$+9$	$+16$	$+25$

(The $+$ signs are there to remind you that the values are positive. They are not essential, and the table would still be quite correct if they were omitted.)

   Now draw the graph. Again the y values go from 0 to 25.

Notice that the y axis is a *line of symmetry* of this curve; this symmetry can often help in drawing a neat curve.

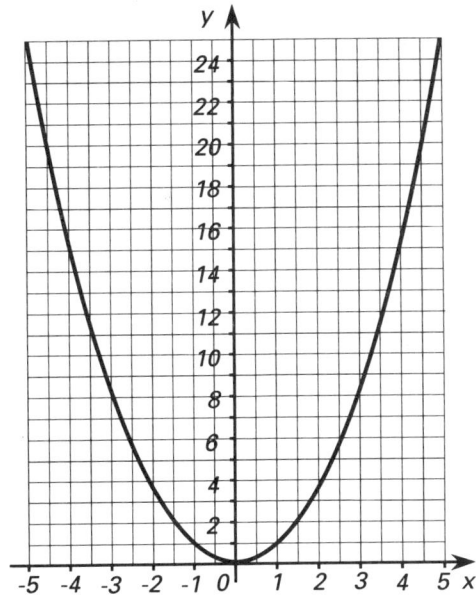

Example 13

Draw the graph of $y = 2x^2 - 5$, for x values in the range -3 to $+4$.

The table will be:

x	-3	-2	-1	0	1	2	3	4
x^2	$+9$	$+4$	$+1$	0	$+1$	$+4$	$+9$	$+16$
$2x^2$ -5	18 -5	8 -5	2 -5	0 -5	2 -5	8 -5	18 -5	32 -5
$y = 2x^2 - 5$	13	3	-3	-5	-3	3	13	27

The y values range from -5 to $+27$.

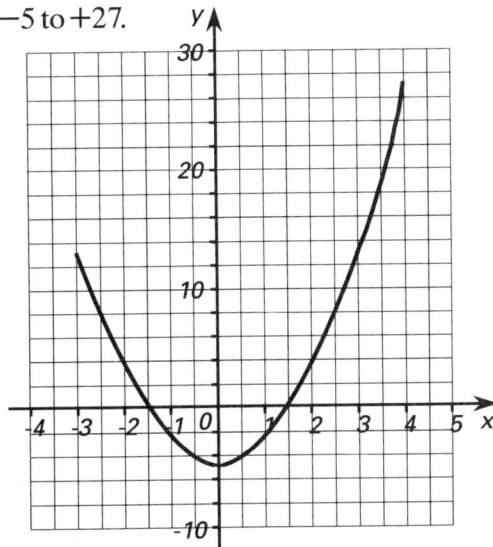

EXERCISE 7.6

Draw the graphs of these equations in each question, by making up a table and using the range of x values given.

1 $y = x^2 + 2$, x values from 0 to 6.

2 $y = x^2 - 7$, x values from -3 to 4.

3 $y = 3x^2$, x values from -3 to 3.

4 $y = 4x^2 - 10$, x values from -3 to 4.

5 $y = x^2 + 2x + 3$, x values from -3 to 3.

Gradient

The road signs warning of steep hills often have something like 'Steep hill – gradient $\frac{1}{7}$' written on them. For that stretch of the road, this means that it will rise one metre for every seven metres along.

Similarly, a gradient of '1 in 4' means that the road will go up by one metre for every four metres along.

In mathematics we use the term **gradient** in a similar way to tell us the steepness of a straight line. Although some road signs use a percentage, in mathematics we usually represent the gradient by a fraction, or a decimal. Instead of '1 in 4' we would write $\frac{1}{4}$ or 0.25, which means one metre *vertically* for every four metres *horizontally*, not four metres *along the road*.

Example 14

What is the gradient of the line in the diagram?

Choosing two points on the line, we can see that it goes up 2 units for every 3 units along.

The gradient is therefore $\frac{2}{3}$.

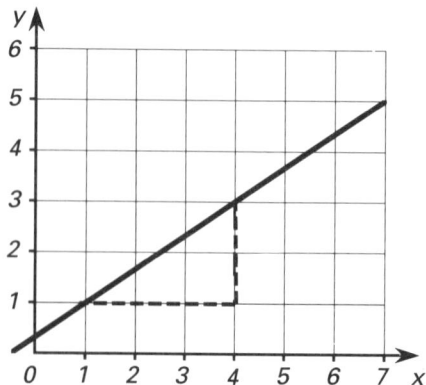

Example 15

Find the gradient of the lines which pass through (*a*) P and Q (*b*) R and S.

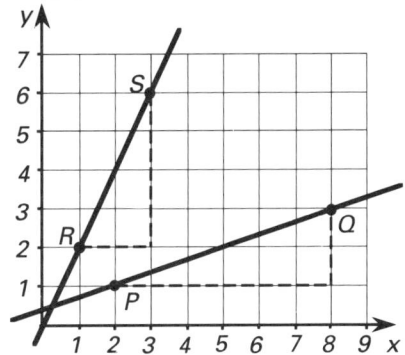

(*a*) The line from P to Q goes up 2 for every 6 along. Its gradient is
therefore $\frac{2}{6}$ or $\frac{1}{3}$.

(*b*) The line from R to S goes up 4 for every 2 along. Its gradient is
therefore $\frac{4}{2}$ or 2.

Example 16

What are the gradients of the lines AB and CD?

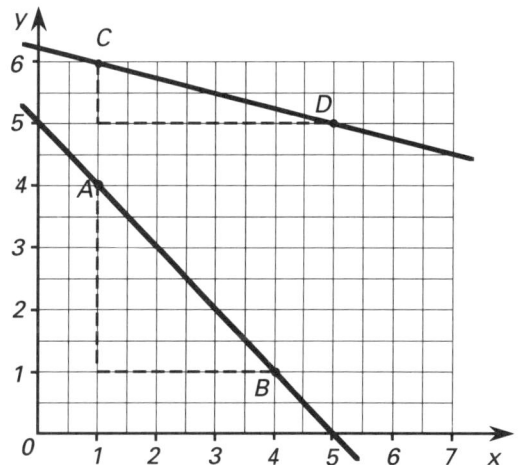

AB goes *down* 3 units for every 3 units along. Its gradient is therefore $\frac{-3}{3}$
or -1.

CD also goes down by 1 unit for every 4 units along, so its gradient is $-\frac{1}{4}$.

As you move from left to right, any line which goes *up* will have a *positive* gradient, and any line which goes *down* will have a *negative* gradient.

Because we divide the distance up (or down) by the distance across in order to give us the gradient, we can also say that the distance up (or down) *for every one unit across* is the gradient. So a gradient of $\frac{1}{4}$ can be looked upon as '1 up for every 4 along' or '$\frac{1}{4}$ up for every 1 along'.

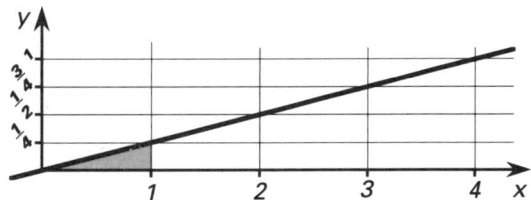

EXERCISE 7.7

1 Work out the gradient of each of the lines in the diagram.

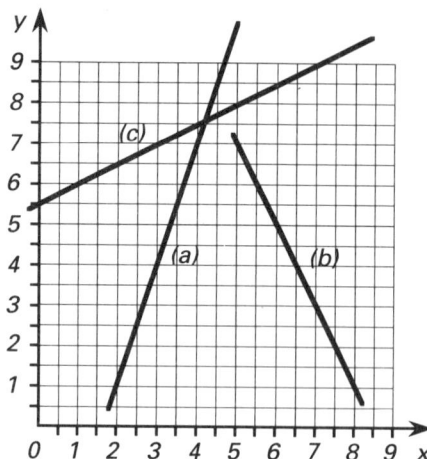

2 Work out the gradient of the line joining (*a*) (1,4) to (6,9), (*b*) (1,4) to (9,6).

3 (*a*) What is the gradient of the line which joins (3,0) to (9,2)?
(*b*) By mistake, Jim joined (3,0) to (2,9). What is the gradient of Jim's line?

4 Join the point (−3,6) to the origin (0,0). What is the gradient of this line?

5 (*a*) Plot the points K(−2,2), L(−3,−4) and M(0,−2) and join them to form triangle KLM.
(*b*) Work out the gradients of the three sides KL, LM and KM.

6 A line is drawn through the point (2,3). It has a gradient of 0. Write down the coordinates of three other points on the line. What can you say about the coordinates of any point on this line?

7 What is the gradient of the line joining (*a*) (0,8) to (8,0) (*b*) (0,5) to (5,0) (*c*) (10,0) to (0,10) (*d*) (*h*,0) to (0,*h*), where *h* is any number you choose?

8 My lean-to greenhouse roof slopes down from the wall of the house, as in the diagram. The four corners of the end of the greenhouse are at the points (0,0), (6,0), (0,9) and (6,6).

(*a*) What is the gradient of the strengthening bar which joins
(0,0) to (6,6)?

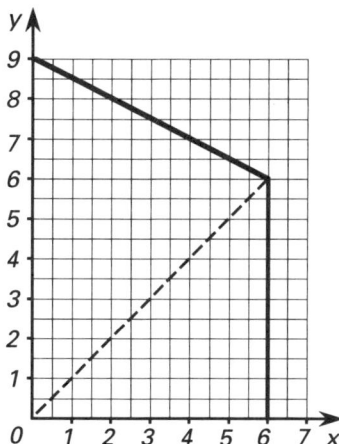

(*b*) What is the gradient of the roof, i.e. from (0,9) to (6,6)?

(*c*) Another bar joins (0,9) to (6,0). What is its gradient?

9 (*a*) A line which has a gradient of $\frac{1}{2}$ passes through the point
(0,3). Draw this line on a grid (take the x axis from
-4 to $+5$).

(*b*) Another line, with gradient -1, passes through the origin
(0,0). Draw this line on the same grid.

(*c*) What are the coordinates of the point where the two lines
meet?

10 What can you say about the gradient of the line joining the
points (4,1) and (4,6)?

Solving simultaneous equations graphically

Example 17

Find the coordinates of the point where the lines

$y = x - 4$ and $y = \frac{x}{2}$ meet.

(*a*) Make up a table for each line, choosing any range of x values.

x	0	1	2	3	4	5	6
$y = x - 4$	-4	-3	-2	-1	0	1	2

x	0	1	2	3	4	5	6
$y=\dfrac{x}{2}$	0	0.5	1	1.5	2	2.5	3

(*b*) Now draw these two lines.

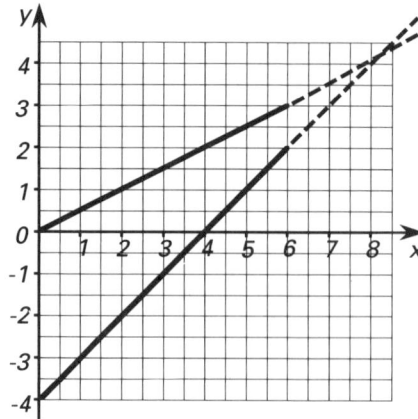

(*c*) Clearly the lines do not quite meet. By extending them, however, they will meet at the point (8,4).

If we extend the *x* value up to 8 in each table, the *y* value will be 4 in each case.

In algebraic terms, we found that the *x* and *y* values which satisfy the two equations $y=x-4$ and $y=\dfrac{x}{2}$ *simultaneously* are $x=8$ and $y=4$.

Example 18

Find the coordinates of the point of intersection of the lines $y=3x-4$ and $y=8-x$.

(*a*) Make up the two tables (Remember: $8-x=8+(-x)$.)

x	0	1	2	3	4	5	6
$3x$ -4	0 -4	3 -4	6 -4	9 -4	12 -4	15 -4	18 -4
$y=3x-4$	-4	-1	2	5	8	11	14

x	0	1	2	3	4	5	6
8 $-x$	8 0	8 -1	8 -2	8 -3	8 -4	8 -5	8 -6
$y=8-x$	8	7	6	5	4	3	2

(*b*) Draw the lines.

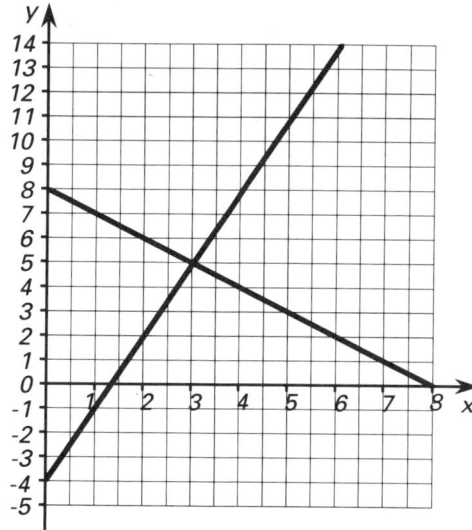

(*c*) The point where the lines meet is (3,5).

   We could have obtained the answer from the two tables. However, the answer seldom occurs in the tables of values, since (i) the range that we have chosen may not cover the point where the lines meet, and (ii) the point where the lines meet may not have whole number coordinates. We may therefore have to estimate the coordinates, and hence we need to be as accurate as we can when plotting the points and drawing the lines.

EXERCISE **7.8**

For each question, find the coordinates of the point where the two lines meet. In some of the simple cases it may not be necessary to make up a table in order to draw the line. You should, however, work out *three* points on each of these lines, before drawing them.

1 $y = x$ and $y = 6 - x$

2 $y = 10 - x$ and $y = 2x - 5$

3 $y = \dfrac{x}{3}$ and $y = x - 6$

4 $y = \dfrac{x}{2} + 3$ and $y = 2x - 3$

5 $y = -x$ and $y = x + 8$

6 $y = -\dfrac{x}{2} - 2$ and $y = x - 7$

7 $y = 2x + 1$ and $y = 5$

8 $y = 3x + 4$ and $y = \dfrac{x}{2} - 2$

9 $y = x - 1$ and $y = 12 - 2x$

10 $y = x$ and $y = 2x$

8. Communication: graphics (2)

Bisecting a line

In Book 3X, Chapter 7 we used our drawing instruments to draw accurately triangles of different sizes. Drawing instruments can also be used to help us with some geometry. The word **bisect** means divide into two equal parts.

A ———————————————————— B

We want to bisect the line AB. How do we do it?

Method 1: Using a ruler, measure the line, divide this measurement into two, measure half way from one of the ends and place a mark at the centre.

Method 2: This method uses a pair of compasses and has the advantage of requiring no calculations.

Extend your pair of compasses so they are more than half the length of the line. Place the point of the compass on A, and draw two long arcs: one above and one below the line.

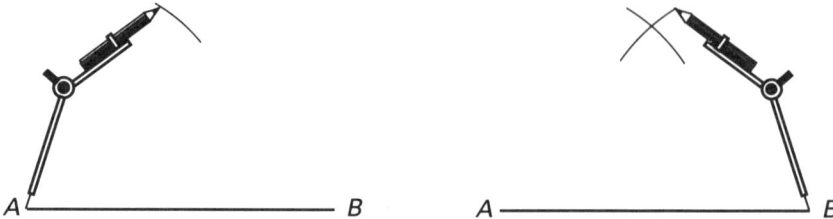

Then, without altering the angle of the compass, place the point of your compass on point B, and draw another two long arcs: one above and one below the line. The arcs should cross as shown.

Join the crosses with a line: this line should bisect AB, and is called the **bisector** of AB. You can check your result by measuring the two halves.

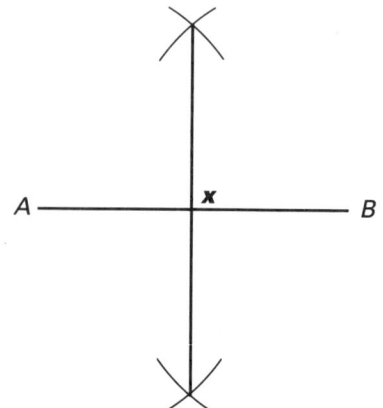

Now draw a line in your book and bisect it yourself. Check you have bisected it correctly by measurement. Measure the angle marked as x on the diagram. What do you find? If you have drawn the bisection accurately the angle x should be about 90°. Not only has this method bisected the line for us, but it has also given us a right angle.

Note: You do not need to rub out the pencil marks you have made. These show clearly which method you have used, and can help in checking your work.

EXERCISE 8.1

Whenever you need to bisect a line use your pair of compasses only.

1 Draw lines with these lengths and bisect them: (*a*) 8 cm
(*b*) 10 cm (*c*) 7 cm (*d*) 11.5 cm.

Draw these triangles accurately, bisecting the side BC in every case:

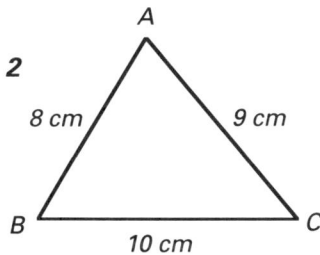

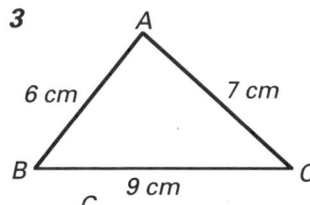

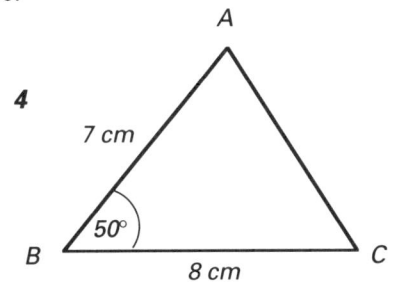

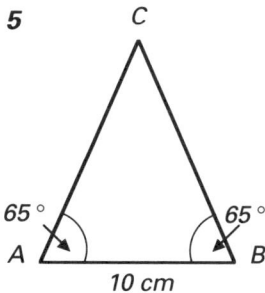

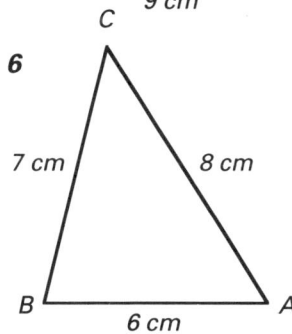

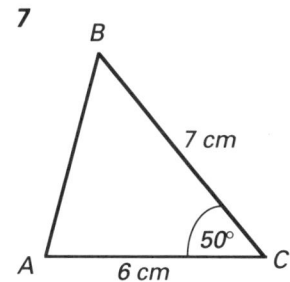

2

Triangle: AB = 8 cm, AC = 9 cm, BC = 10 cm

3

Triangle: AB = 6 cm, AC = 7 cm, BC = 9 cm

4

Triangle: AB = 7 cm, angle B = 50°, BC = 8 cm

5

Triangle: angle A = 65°, angle B = 65°, AB = 10 cm

6

Triangle: CB = 7 cm, CA = 8 cm, BA = 6 cm

7

Triangle: BC = 7 cm, angle C = 50°, AC = 6 cm

8 Draw this triangle accurately, then bisect the line BC. What can you say about the bisector of BC? Can you explain it?

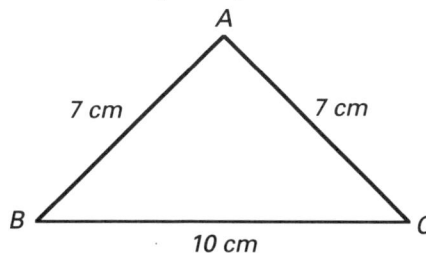

Triangle: AB = 7 cm, AC = 7 cm, BC = 10 cm

Bisecting an angle

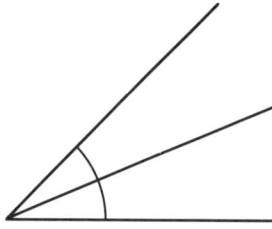

Method 1: Using a protractor, measure the angle, divide the measurement by two, measure this new angle and draw in a line to mark the bisection.

Method 2: This method uses a pair of compasses and requires no calculations.

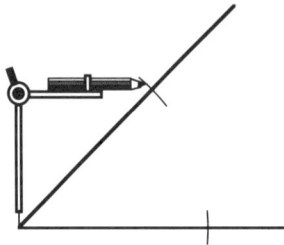

Place the point of the compass on the intersection of the two lines. Keeping the compass at the same extension draw two arcs, one on each line.

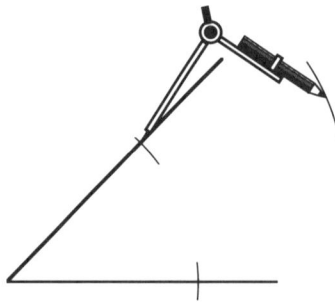

Next place the compass on the first arc you have just drawn where it crosses the line, and add another curve in the position shown. Repeat this for the other arc.

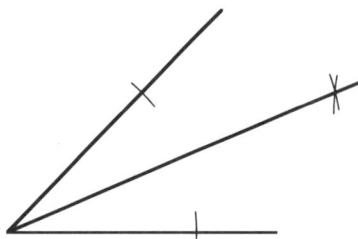

Join the cross to the intersection point: this line will bisect the angle, and is called the **bisector**. We can bisect any angle using this method.

Draw an angle of any size in your book and bisect it using this method. Check you have bisected the angle correctly by measurement using a protractor.

EXERCISE 8.2

Whenever you need to bisect an angle use your pair of compasses only.

1 Draw each of these angles with your protractor, and bisect them using your pair of compasses: (*a*) 40° (*b*) 65° (*c*) 80° (*d*) 120° (*e*) 140°.

Draw these triangles accurately, bisecting each angle A.

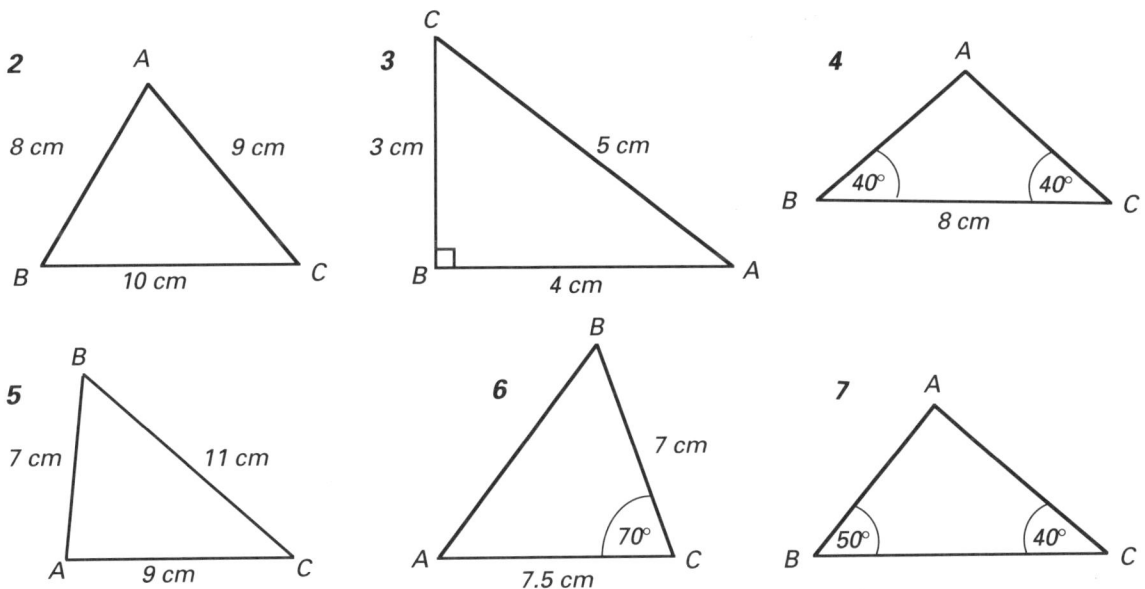

Investigation A

Draw a triangle and bisect each of the sides in turn. What do you notice? Would the same happen for any type of triangle? Try this exercise for an equilateral and an isosceles triangle.

Draw a triangle and bisect each of the angles in turn. What do you notice? Would the same happen for any type of triangle? Try this exercise for an equilateral and an isosceles triangle.

Drawing angles

We can also use a pair of compasses to construct certain angles we may
want regularly, particularly the 90° angle, and the 60° angle.

Example 1

Draw a 90° angle at a point A on a line.

Keeping the compasses extended to the same length, and the point on A,
draw two arcs either side of A.

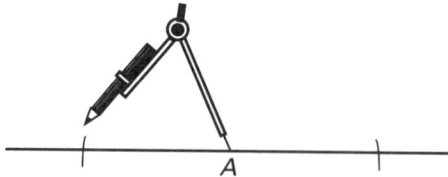

You can now use these two marks to construct a bisector as we have done
before as we know already that this will give us a 90° angle.

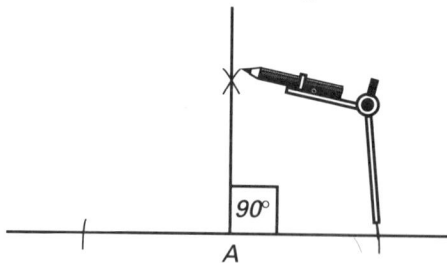

Example 2

Draw a 60° angle at a point A on a line.

Keeping the compasses extended to the same length, and the point on A,
draw two arcs as shown.

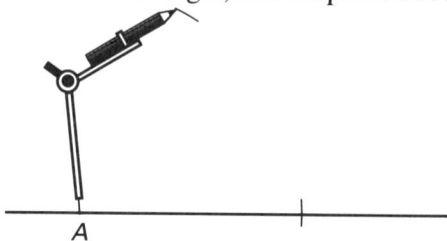

Placing the compass on one of the arcs, draw a third arc as shown to
make a cross. Joining the cross to the point A with a line will give you the
60° angle.

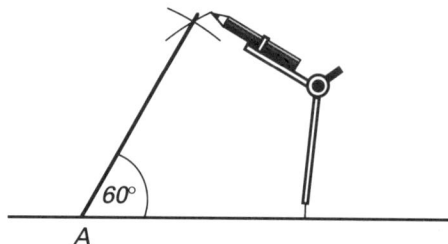

EXERCISE **8.3**

Using compasses only, draw the following angles.

1 90°   **2** 60°

3 Using compasses only, draw a 90° angle and bisect it to get a 45° angle.

4 Using compasses only, draw a 60° angle and bisect it to get a 30° angle.

5 Using the same methods as in questions 3 and 4, draw a 15° angle, and a $22\frac{1}{2}°$ angle.

EXERCISE **8.4**

Draw the following shapes accurately using only a pair of compasses and a ruler.

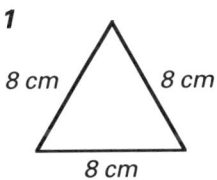

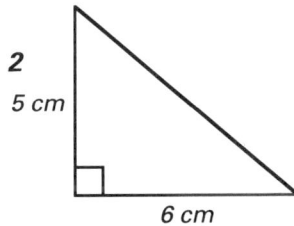

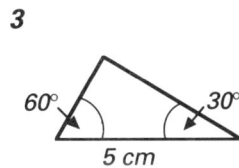

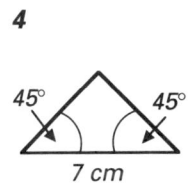

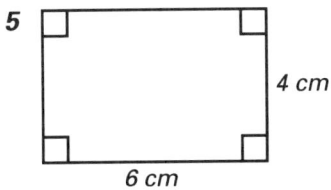

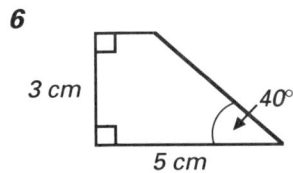

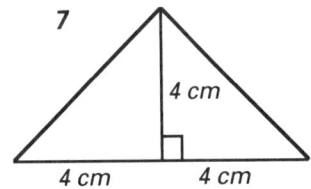

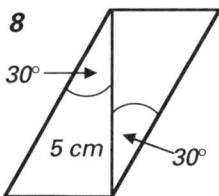

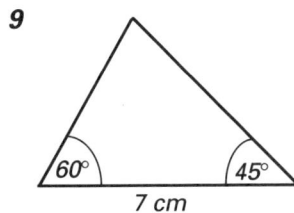

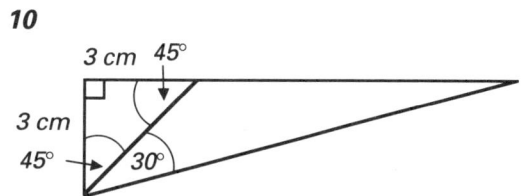

1

8 cm   8 cm

8 cm

2

5 cm

6 cm

3

60°   30°

5 cm

4

45°   45°

7 cm

5

4 cm

6 cm

6

3 cm   40°

5 cm

7

4 cm

4 cm   4 cm

8

30°

5 cm   30°

9

60°   45°

7 cm

10

3 cm   45°

3 cm

45°   30°

Loci

A locus is a series of points which satisfy a rule. In particular, when we trace the path made by joining this series of points, this path is called the **locus** of the points. The plural of locus is loci.

Example 3

The locus of a golf ball is something like this.

EXERCISE 8.5

Describe or sketch the paths of the following:

1 The movement of the end of a minute hand of a clock through half an hour.

2 The path of a child on a slide.

3 The movement of a man's hand while throwing a dart.

4 A car overtaking a stationary van.

5 The movement of the midpoint of a skipping rope when in use.

6 The gear stick of a car when the gears are changed.

7 A rugby ball when it is kicked between the posts.

8 A netball when it is thrown through the net.

9 The path made by a swimmer jumping off a diving board.

10 The end of a windscreen wiper of a car while in motion.

11 A ball being bounced off the floor.

12 The movement of the head of a jogger.

13 The path of a lift as it rises through a building.

14 The head of a hammer as it is used to strike a nail.

15 The path made by the head of a jack-in-the-box as it is released.

Example 4

A goat is tethered at one corner of a field by a rope 10 m long. Shade in the region on a diagram within which the goat can wander.

The locus of the boundary for the goat is actually an arc across the field, and the region to be shaded is that between the arc and the sides of the field.

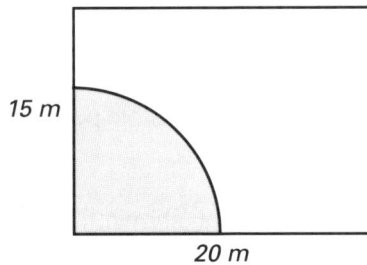

15 m

20 m

EXERCISE 8.6

1 Draw a scale plan of a table 120 cm long and 80 cm wide. Draw the locus of a fly which crawls around the table keeping 20 cm from the edge.

2 A pencil is fastened to a peg by a length of string. The pencil is moved around the peg while keeping the string tight, and the string is gradually wound in. Describe the locus of the pencil.

3 Draw a straight line across your page. Describe the locus of the points which are (*a*) 2 cm (*b*) 3 cm from the line.

4 Place two points, A and B, on your page. Describe the locus of points which are (*a*) 2 cm from A (*b*) 3 cm from B.

5 Place two points on your page. Describe the locus of points that are 3 cm from one of the points, and also 4 cm from the other point.

EXERCISE 8.7

1 (*a*) Describe the locus of a ball which is rolled towards D, so that it is at equal distances from the two walls AD and DC.

(*b*) A light at B illuminates all points no greater than 6 m from the light. Shade the area lit up.

(*c*) Describe the locus of points which are at equal distances from AD and BC.

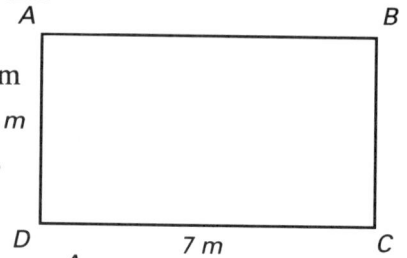

2 (*a*) Describe the locus of all points 5 cm from A.

(*b*) Describe the locus of all points 3 cm from B.

(*c*) Shade the region where all points are within 5 cm from A *and* 3 cm from B.

(*d*) Describe the locus of all points which are at equal distances from A and C.

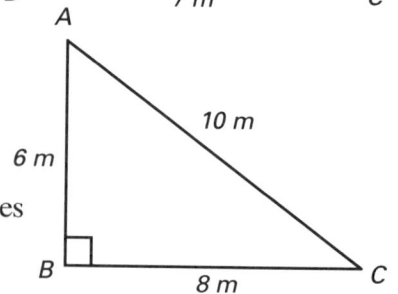

3 Two houses are 30 m apart. The junction of the water main between the two houses lies 18 m from A and 16 m from B. Draw a scale diagram and mark on it the two possible positions of the water main junction.

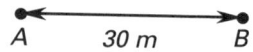

4 Three television transmitters are positioned as shown. The transmitters have a range as follows: A, 50 miles; B, 60 miles; and C, 40 miles. Shade the part of the diagram where signals from all three transmitters can be received.

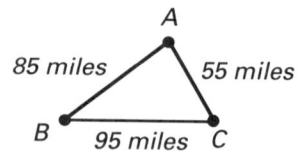

5 Two moles dig their way under a garden. One keeps 5 feet away from edge DC, while the other starts from B and keeps an equal distance from AB and BC. Find out where their two tunnels will meet.

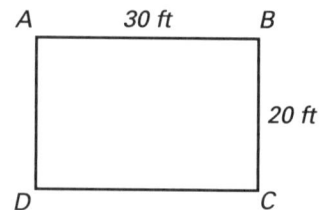

Investigation B

A ladder slides down a wall as shown.
On graph paper draw in some other positions of the ladder. The midpoints of the ladder in some positions are marked. Join up all the midpoints you have drawn and find the locus of these midpoints. Extend the problem to find the locus of other points on the ladder.

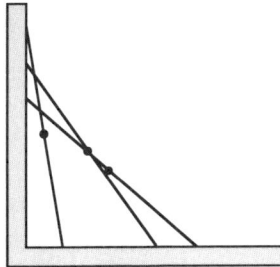

Investigation C

A further extension to the problem in Investigation F is to consider ladders as shown, between two lines which are not at right angles. Again find the locus of the midpoints, or any other points on the ladder.

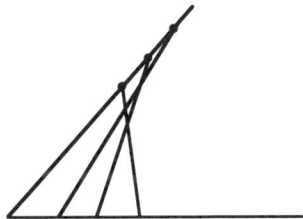

Investigation D

On a board place two pins 9 cm apart, and connect the two pins with a string 12 cm long. Keeping a pencil in contact with the paper with the string taut, move the pencil around the points. What shape can you draw? What shape is produced when you reduce the length of the string, or move the pins?

9. Communication: geometry

Congruence

Which of the shapes P, Q, R and S are exactly the same as shape A?

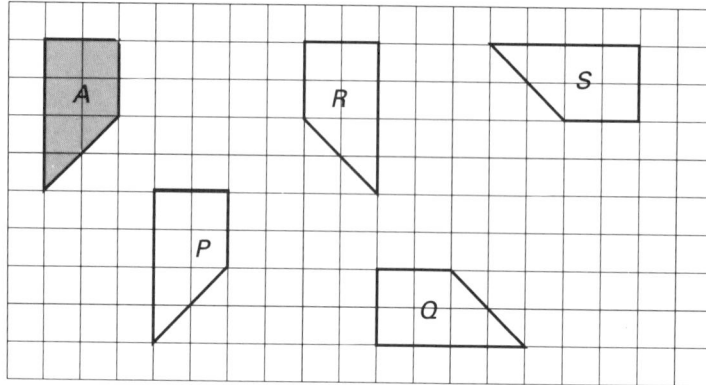

Only P is exactly the same as A, if we mean the same way up as well as being the same shape and size.

However, all four are exactly the same as A, if we mean by 'exactly the same' that A, cut out of paper, could fit exactly over P, Q, R or S.

Shapes are **congruent** when they are exactly alike in all respects, so that they will fit exactly on top of one another with no overlap. (One or more of the shapes may have to be turned over, but that doesn't matter, the shapes will still be congruent.)

So in the diagram above, the five shapes A, P, Q, R and S are all congruent with one another.

EXERCISE **9.1**

1 (*a*) Which shapes are congruent with shape T?

(*b*) Draw two further shapes, congruent with T.

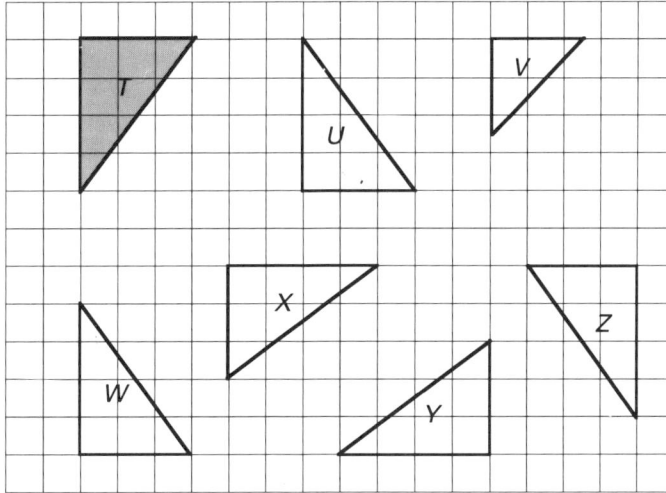

2 Divide this L-shape into four congruent L-shapes.

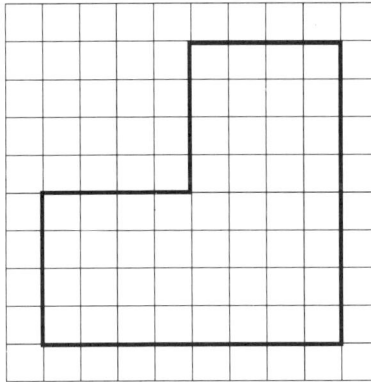

3 (*a*) Draw a grid with this shape on it.

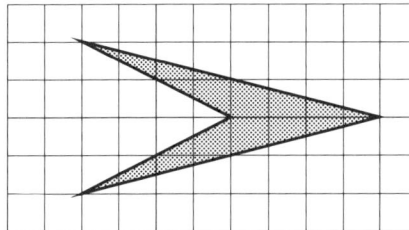

(*b*) Now draw two shapes which are congruent with this shape, but which point in different directions.

4 One side AB of a triangle ABC, is shown in the diagram. There are four possible positions of C, so that triangle ABC is conguent with triangle DEF.

(*a*) What are the coordinates of the possible positions of C?

(*b*) State the coordinates of the three corners of another triangle which is congruent with triangle DEF.

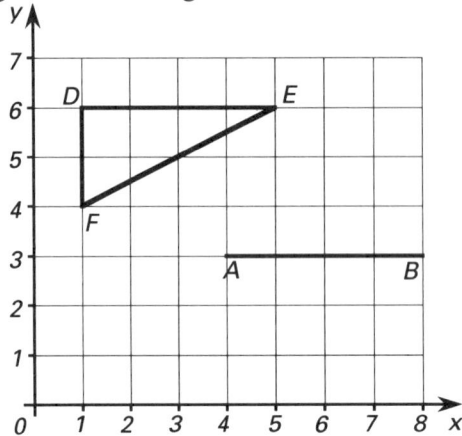

5 Triangle T has sides of length 3 cm, 4 cm and 5 cm. By shading on squared paper, show how to join the equal sides of two triangles, congruent with T, to form:

(*a*) a rectangle 3 cm by 4 cm

(*b*) a triangle with sides of 5 cm, 5 cm and 6 cm

(*c*) a triangle with sides of 5 cm, 5 cm and 8 cm

(*d*) a parallelogram with sides 3 cm and 5 cm

(*e*) a parallelogram with sides 4 cm and 5 cm

(*f*) a 'kite' with two sides of 3 cm and two of 4 cm.

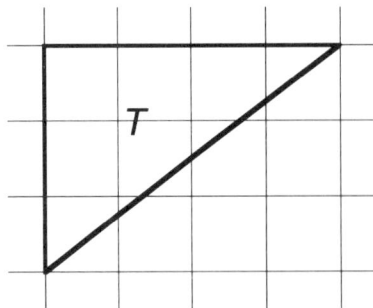

Investigation A

In question 5, you drew all six possible combinations where the equal sides of two congruent triangles could be joined to each other.

(*a*) Now repeat, using three congruent triangles. Try to arrange your drawings into some sort of order.

(*b*) Repeat, using four congruent triangles.

Investigation B

(*a*) Draw any quadrilateral on a piece of card and cut it out.

(*b*) Place the card shape on a piece of paper, and draw round it.

(*c*) Rotate the card a half-turn about the middle point of one of the sides, and draw round it. (You now have two congruent shapes.)

(*d*) Repeat the operation, using the middle point of another side.

(*e*) Can you fill the page, leaving no gaps?

Try the whole process again, this time with a pentagon. Are there gaps this time? Can you choose another pentagon, or another polygon, which will leave no gaps? What about a regular pentagon, or hexagon?

Investigation C

(*a*) Are all triangles with sides 5 cm, 6 cm and 7 cm congruent with each other?

(*b*) Are all quadrilaterals with sides 5 cm, 6 cm, 7 cm and 8 cm congruent with each other?

Similarity

In everyday conversation when we talk about two objects which are similar, we usually mean that there are some aspects which are the same in both objects, but there are, also, some differences.

When we talk about similar objects in mathematics, we mean something much more precise.

If one object is an exact scaled up (or scaled down) version of the other, then the objects are said to be **similar**.

Let us begin by looking at some simple shapes.

Example 1

Which of these triangles are similar to triangle T?

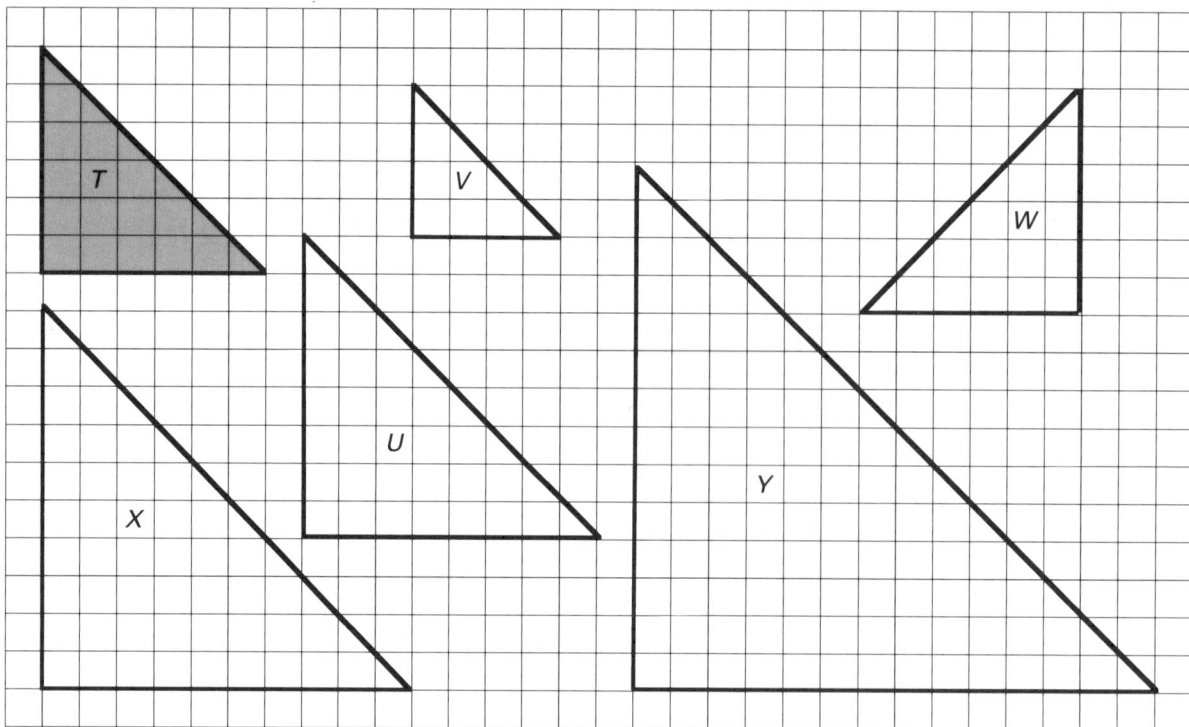

In fact, all of them are similar. As with congruence, it doesn't matter
whether or not the shape has to be turned round, or turned over. As long
as one triangle is a scale drawing of the other then they are similar. (It can
help to compare two shapes, if one is redrawn the same way up as the
other.)

Example 2

Draw a shape similar to shape Z, but with its sides twice as long.

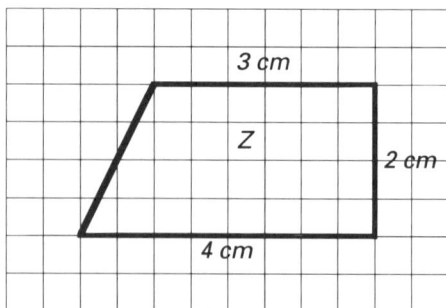

All lengths have to be doubled.

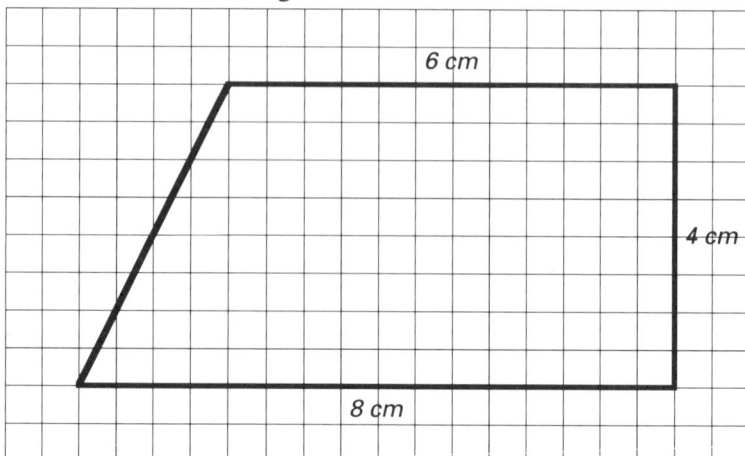

Example 3

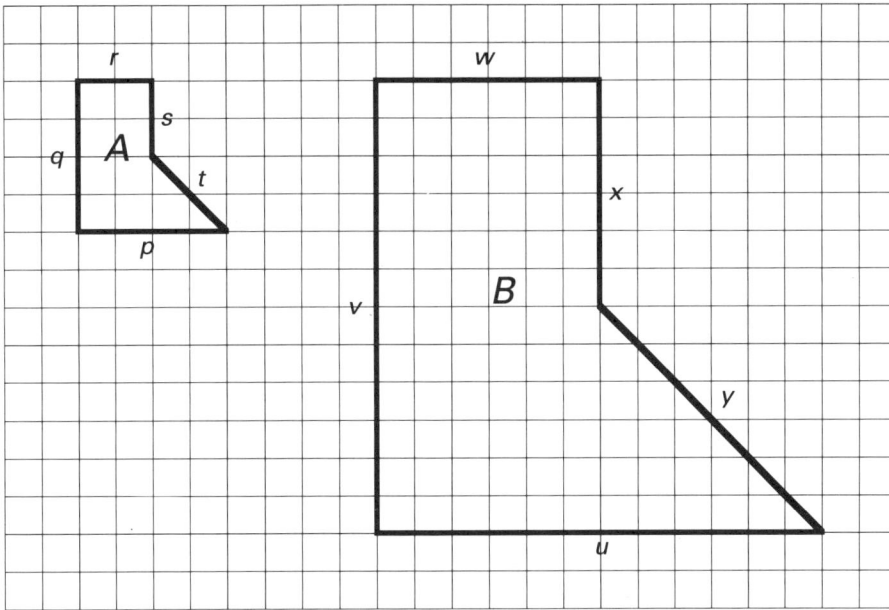

Shapes A and B are similar.

(*a*) Measure the lengths of the sides of each shape.

(*b*) Write down the lengths you have measured in a table.

Shape A	Shape B	Scale factor
$p = 2$ cm	$u = 6$ cm	$6 \div 2 = 3$
$q =$	$v =$	
$r =$	$w =$	
$s =$	$x =$	
$t =$	$y =$	

(*c*) Divide one of the sides in B by the corresponding side in A.

(*d*) Repeat the procedure for all the sides.

(*e*) What do you notice?

This value of 3, which is the same for each corresponding pair of sides, is the **scale factor** of the enlargement.

  (If the corresponding sides of two similar shapes are parallel, then one of the shapes is an **enlargement** of the other.)

EXERCISE **9.2**

1 Which of these shapes are similar to shape P?

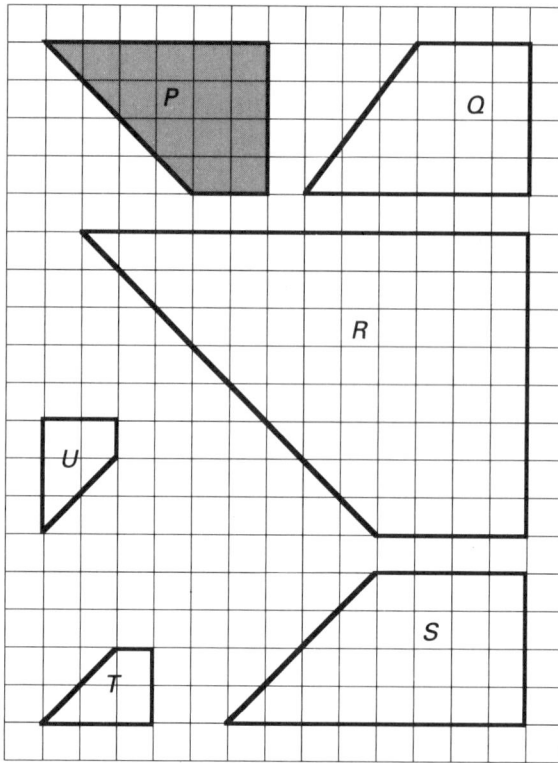

2 Copy and complete the diagram, so that the two shapes are similar to shape X.

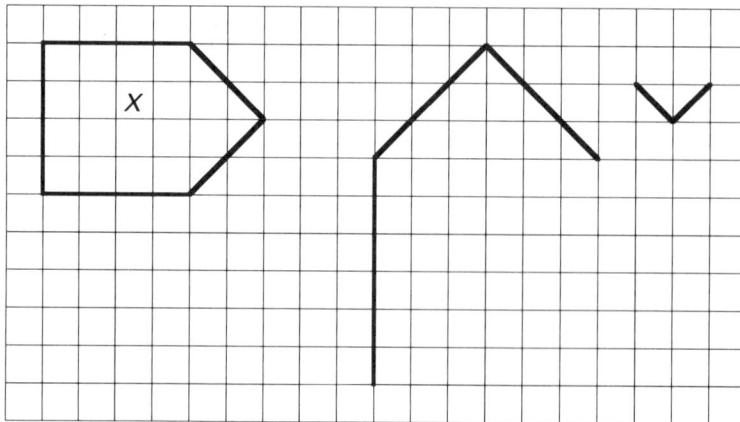

3 A photograph is 12 cm long and 9 cm high. It is enlarged so that the length of the enlargement is 60 cm. How high is the enlargement?

60 cm

12 cm

9 cm   ?

4 (*a*) Draw an enlargement, scale factor 2, of this shape.

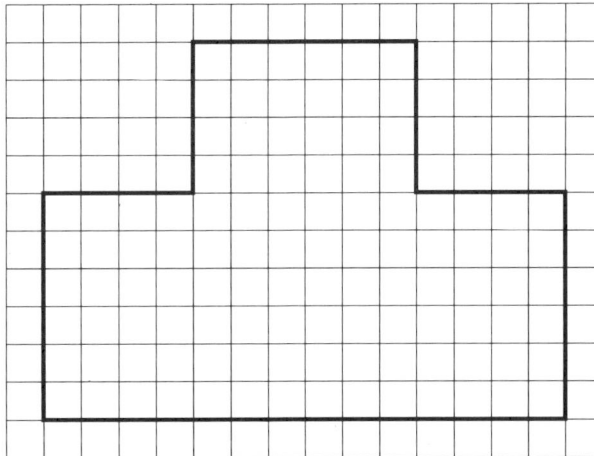

(*b*) Draw another enlargement, scale factor 2, somewhere else on the page.

(*c*) What can you say about the two enlargements?

5 As part of the evidence for an assessment in Art, I need to have a photograph of my painting. The photograph measures 8 cm long by 5 cm wide. If my actual painting is 32 cm long, how wide is it?

8 cm

5 cm

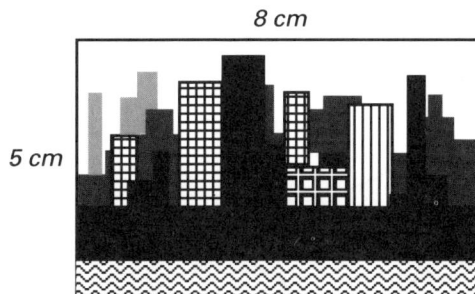

Angles between parallel lines

The two lines PQ and RS are parallel; therefore they point in the same direction, will never meet, and are always the same distance apart. (We often denote parallel lines by putting an 'arrowhead' on each of them, pointing in the same direction.)

If another line XY is drawn which is *not* parallel to these two lines, it must cut each of them.

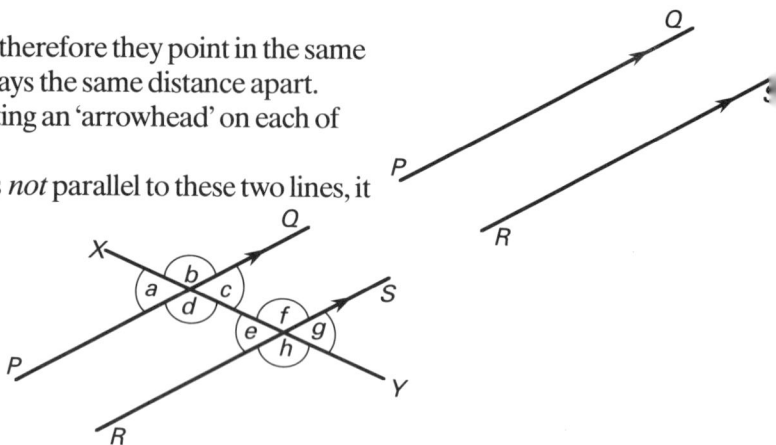

If you measure angles *a* and *e* with your protractor you will find that they are equal. This is because the angle turned through anticlockwise from the direction XY to direction PQ is the same as the angle turned through anticlockwise from XY to RS, since PQ and RS point in the same direction.

The angles *a* and *e* are called **corresponding** angles.

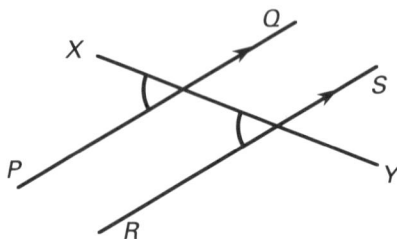

Corresponding angles

Angles *d* and *h* are also corresponding angles, hence they are equal to each other.

Which other pairs of angles are corresponding?

Now look at angles *a*, *b* and *c*. Angles *a* and *b* must add up to 180°, as they are on the line PQ. Also angles *b* and *c* must add up to 180°, as they are on the line XY. If we now subtract *b* from 180° in each case, we are left with *a* and *c* being equal.

Opposite angles

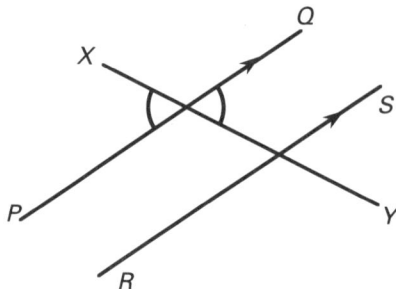

Whenever two lines cut or cross each other (**intersect** is the mathematical term for 'cut' or 'cross'), the **opposite** angles are always equal. So angles a and c are equal, and angles b and d are also equal.

Now look at angles c and e. Because angles a and c are equal (opposite angles) and angles a and e are equal (corresponding angles), then it follows that angles c and e must be equal.

These angles, inside the two parallel lines, but on opposite sides of the line XY, are called **alternate** angles.

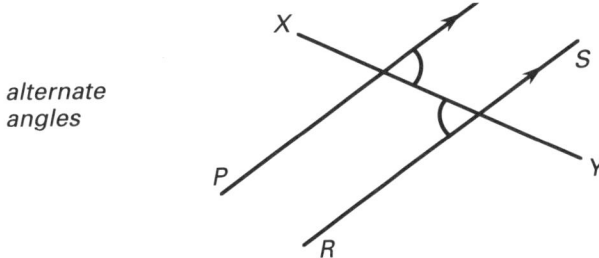

alternate angles

Can you find another pair of alternate angles?

Example 4

(a) Copy the diagram, and mark in *all* the angles which are 50°.

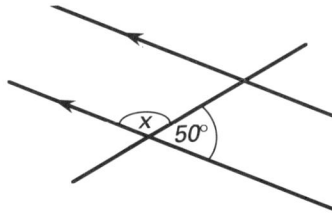

(b) What is angle x? (Remember that angles on a straight line must add up to 180°.)

(c) Mark in *all* the other angles which are equal to angle x.

(a) There are three other angles of 50°.

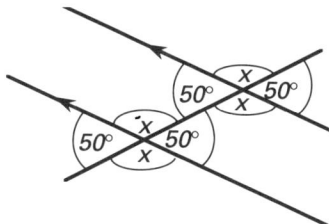

(b) Angle $x = 180° - 50° = 130°$.

(c) The other three angles equal to x ($= 130°$) are marked.

Referring back to the original diagram, we can obtain a further piece of useful information.

As angles *a* and *d* must add up to 180°, and as angle *a* equals angle *e*, then angles *d* and *e* must also add up to 180°.

This is always the case; the two angles between the parallel lines, on the same side of XY, add up to 180°.

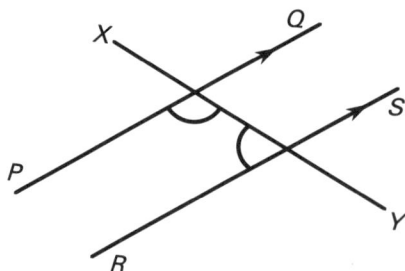

EXERCISE 9.3

In each question work out the angles marked with small letters.

1

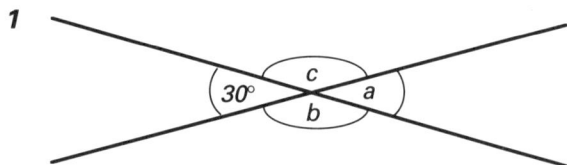

2

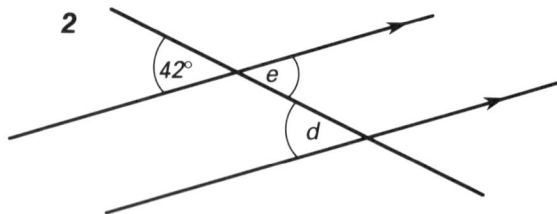

3

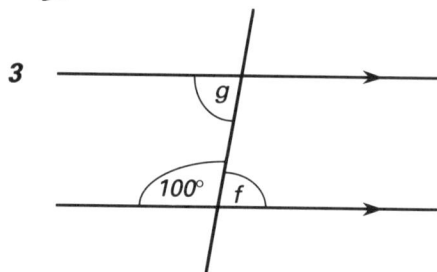

4

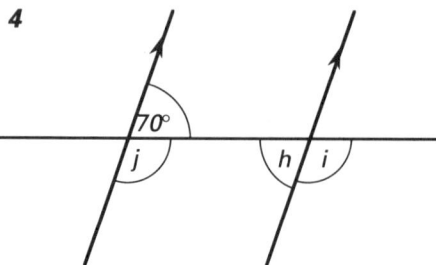

5

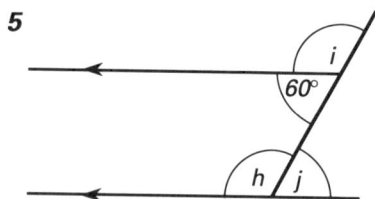

6

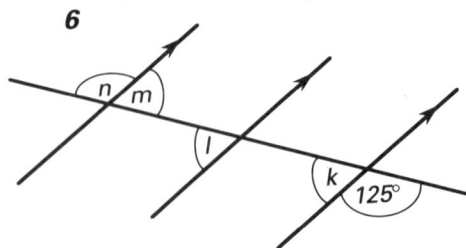

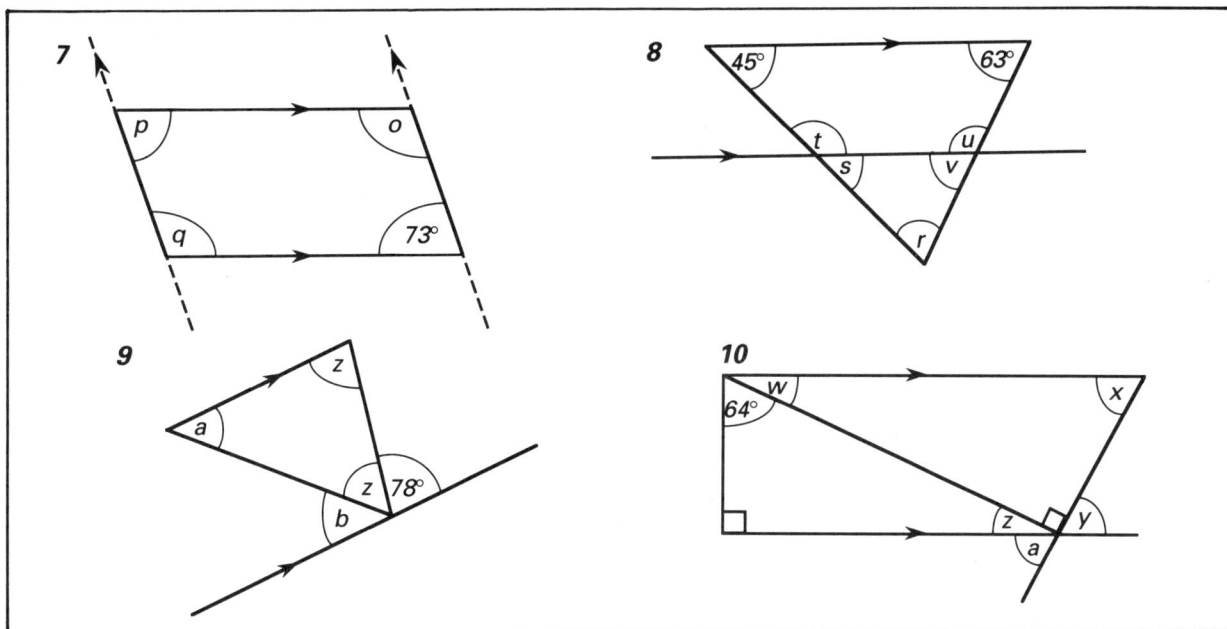

7

8

9

10

Polygons

A **polygon** is an enclosed figure with straight sides. The simplest polygon is a *triangle*.

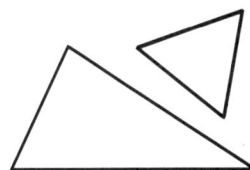

These are examples of polygons with four sides. They are all called *quadrilaterals*.

Here are some polygons with five sides, called *pentagons*.

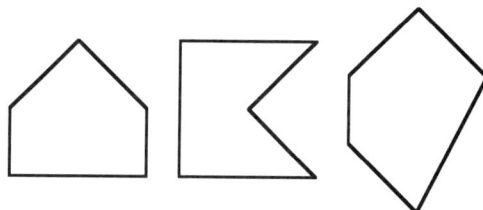

Six-sided polygons are called *hexagons*.

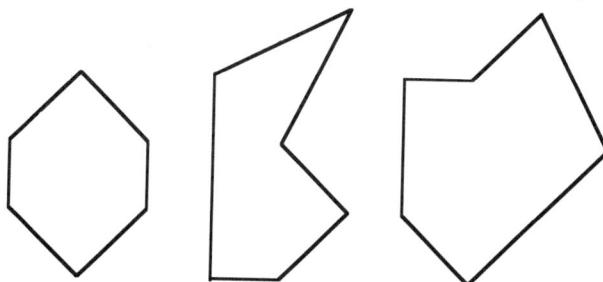

Angles in a polygon

You know that if you measure the three angles in any triangle and add them, the total is always 180°. We use this fact to enable us to investigate the sum of the angles in any polygon.

Investigation D

(*a*) Draw a quadrilateral of any shape you like.

(*b*) Measure the four angles.

(*c*) Add them.

(*d*) Repeat, with another quadrilateral.

(*e*) What do you notice?

(*f*) Try again, with a different quadrilateral.

Example 5

By drawing a diagonal across a quadrilateral, work out the sum of its four angles.

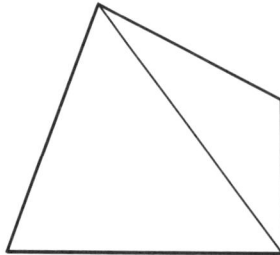

The diagonal divides the quadrilateral into two triangles. The four angles of the quadrilateral add up to the same as the total for the two triangles, which is 360°.

This total will be the same, 360°, for any quadrilateral, because we can always draw a diagonal across a quadrilateral.

Investigation E

(*a*) Draw any pentagon and label the corners A, B, C, D and E.

(*b*) Draw the two diagonal lines from A to C and from A to D.

(*c*) Work out the sum of the angles of your pentagon.

(*d*) Is this total the same for all pentagons?

(*e*) Repeat, this time for a hexagon, drawing all three diagonals from one corner.

Regular polygons

If all of the sides and angles of a polygon are equal, then we call the figure a **regular** polygon.

A regular triangle has a special name; an **equilateral** triangle.

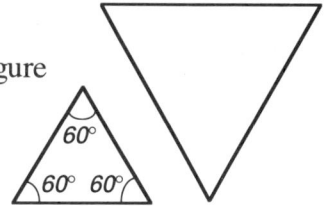

A regular quadrilateral is a **square**.

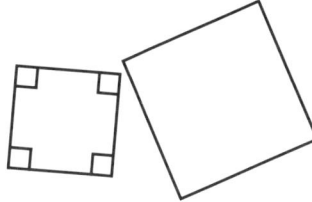

EXERCISE 9.4

Work out the angles represented by the small letters in each of the diagrams.

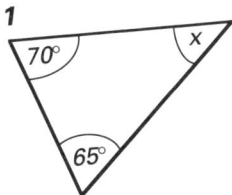

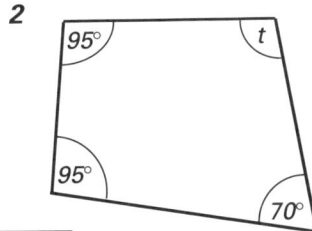

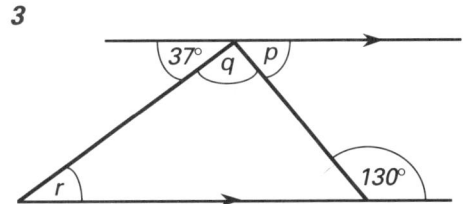

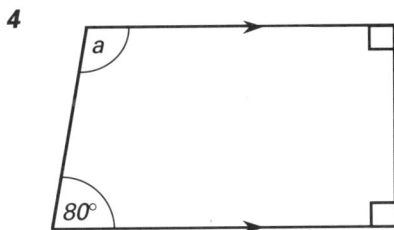

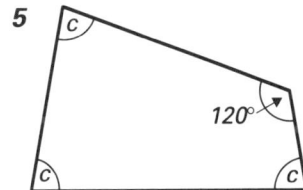

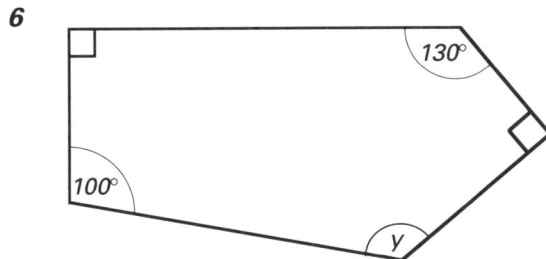

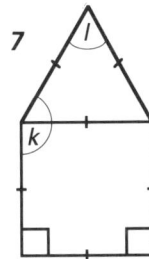

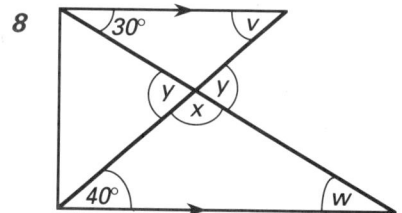

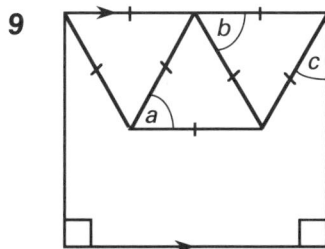

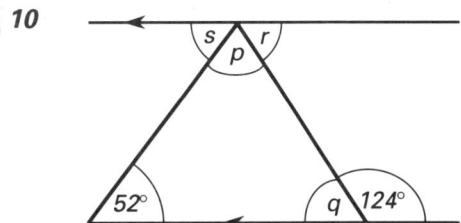

1

70° x

65°

2

95° t

95°

70°

3

37° q p

r

130°

4

a

80°

5

c

120°

c c

6

130°

100°

y

7

l

k

8

30° v

y y

x

40° w

9

b

c

a

10

s r

p

52° q 124°

Example 6

What is the angle at the corners of a regular pentagon?

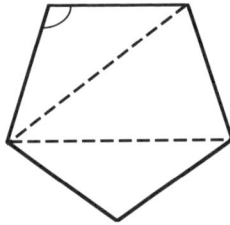

As the sum of the angles in a pentagon is $180° \times 3 = 540°$, then each angle will be $\frac{540°}{5} = 108°$.

We can also approach this problem by looking at the angles made outside the pentagon when each side is extended.

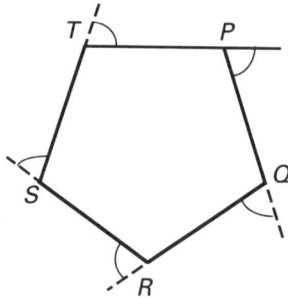

If we imagine that the pentagon is the fence around a field, and that we walk all around the fence PQRST and back to P, we must turn through $360°$; we make one full turn.

So each of the equal exterior angles must be $\frac{360°}{5} = 72°$.

This then enables us to work out that the angle inside each corner must be $180° - 72° = 108°$, as before.

Example 7

What is the angle at a corner of a regular polygon which has twelve sides?

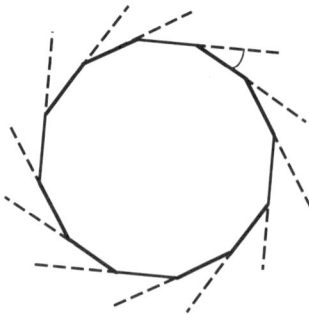

Using the second method in Example 6 above, each exterior angle is $\frac{360°}{12} = 30°$. So the angle at a corner of the polygon is $180° - 30° = 150°$.

EXERCISE **9.5**

Work out the angles in these diagrams.

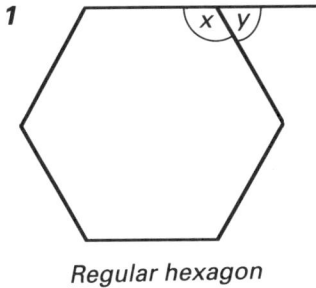

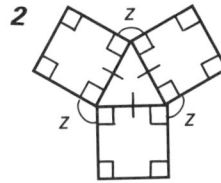

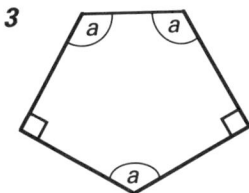

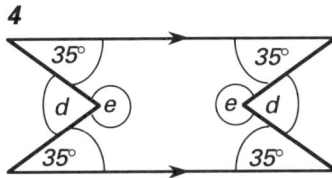

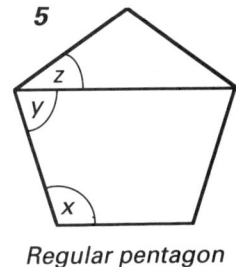

1

Regular hexagon

2

3

4

5

Regular pentagon

Angle in a semicircle

(*a*) Draw a fairly large circle, and label the centre O.
(*b*) Draw a diameter AB.
(*c*) Choose any point C on the circumference of the circle, and draw the lines CA and CB.
(*d*) Measure angle C (i.e. angle ACB).
(*e*) Choose another point (label it D) on the circumference, join it to A and B, and measure angle ADB.
(*f*) Repeat, with another point on the circumference.
(*g*) What can you conclude?

You should have found that, allowing for slight errors in drawing and measuring, the angle formed by joining a point on the circumference to the two ends of a diameter is always 90°.

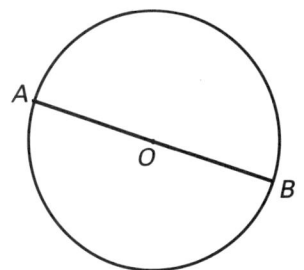

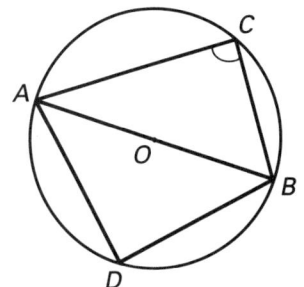

We can summarise this by the statement:

the angle in a semicircle is 90°

Let us see if we can show that the angle will always be 90°, and not 88°, say, or 93°.

In the diagram, XY is a diameter of the circle, centre O, and Z is any point on the circumference.

Let the angles at X and Y be p and q respectively (their size will depend on the position of Z).

Now join Z to O; this gives us two smaller triangles. Each triangle is *isosceles* (i.e. has two sides equal) because the lengths OX, OZ and OY are each equal to the radius of the circle. This means that angle OZX must be p, and angle OZY must be q.

Now the angle in which we are interested, angle XZY, is equal to $p + q$. In the original triangle XYZ, adding all its angles gives $p + q + (p + q)$, which can be written as $(p + q) + (p + q)$. This must be equal to 180° and it follows that $(p + q)$ must be 90°.

So for any point Z on the circumference of the circle, the angle XZY must be 90°.

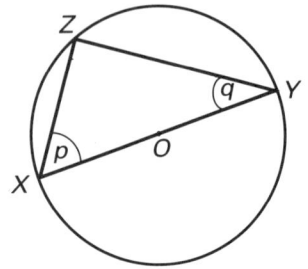

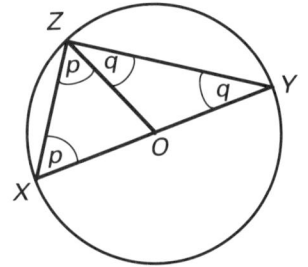

The process of showing that the angle in a semicircle is always 90°, whatever the position of Z, is called a **proof**; we have *proved* the statement that the angle in a semicircle is 90°.

EXERCISE 9.6

In questions 1–5, work out the missing angles. O is the centre of each circle.

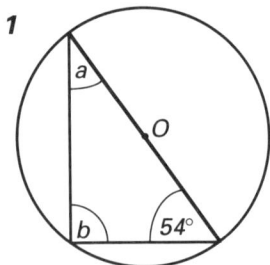

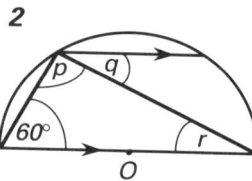

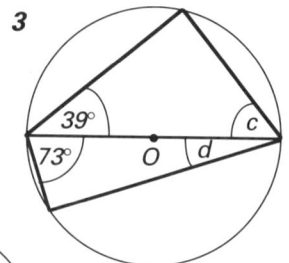

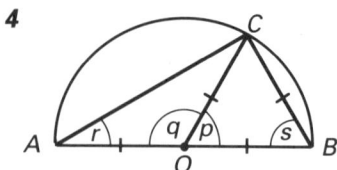

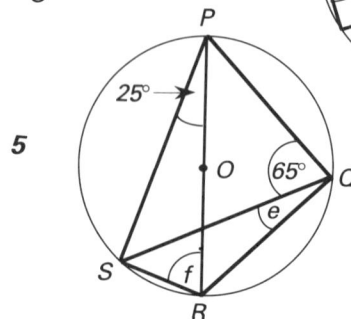

1

2

3

4

5

Angle between tangent and radius

A **tangent** is a line which just touches a circle at one point only.

A table top, for example, is a tangent to a round can of coke or baked beans lying on the table.

If we draw the radius from the centre to the point where the tangent touches the circle, it is clear that the angle between the tangent and the radius is 90°, since we can approach the circle from either side of the tangent, and the two equal angles must add up to 180°.

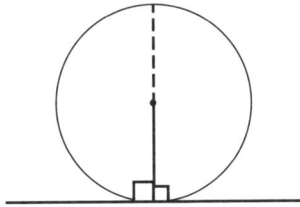

The diagram is symmetrical about the radius. This radius is a mirror line; a reflection of one half of the diagram, using this line as a mirror, forms the other half of the diagram.

Investigation F

Draw a circle, and from a point T outside it, draw as accurately as you can the two possible tangents to the circle.

For each of the tangents, join the point where it touches the circle to the centre O.

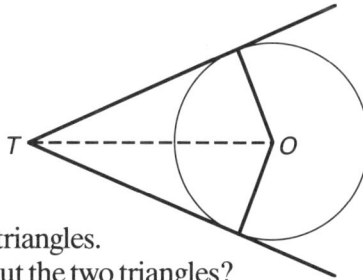

Join T to O to form two triangles.

What can you say about the two triangles?

Repeat for various points T.

Investigation G

Draw a circle, centre O, and mark a point T outside the circle. Mark X as the midpoint of OT.

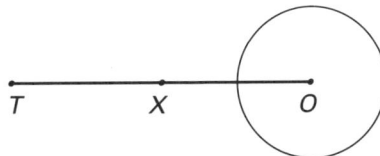

Draw a new circle, with X as centre, and radius XT, so that this circle passes through O and T and cuts the first circle at points A and B. Why are TA and TB both tangents to the original circle?

EXERCISE 9.7

Work out the angles in each of these diagrams. O is the centre of each circle.

1

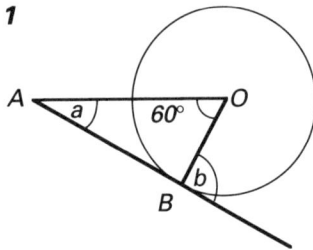

2

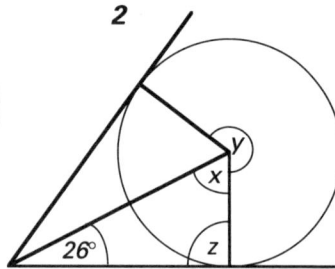

3

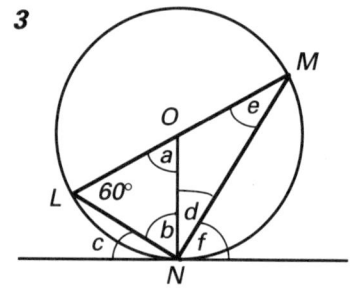

4

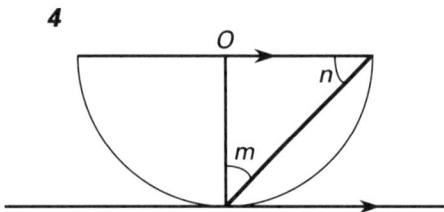

5

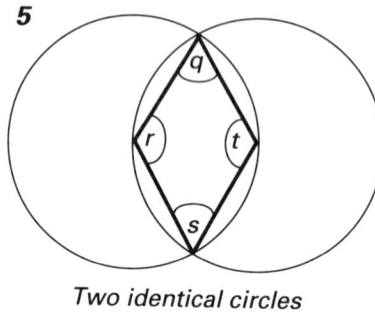

Two identical circles

6

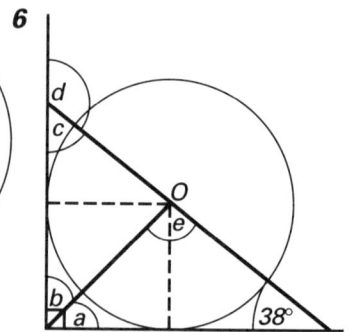

7

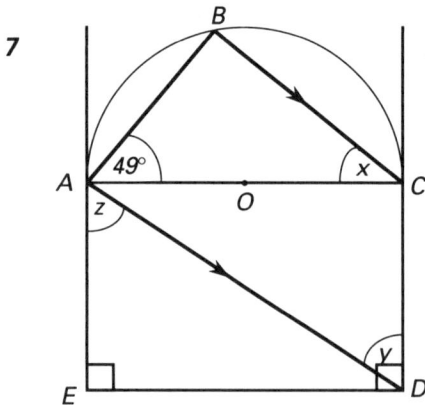

8

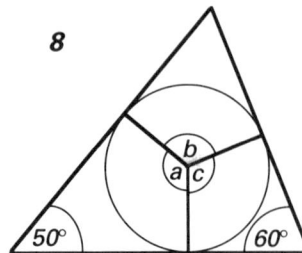

9

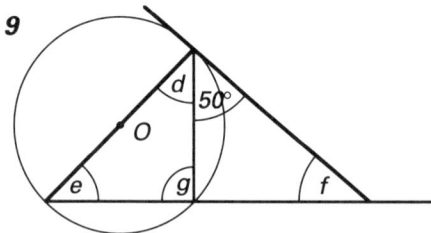

10

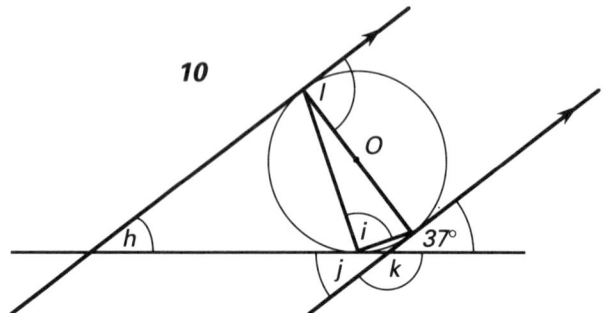

10. Communication: statistics

Histograms

In your previous work, not only in mathematics but in other subject areas, you must have met bar charts or block graphs. Many newspapers, magazines and advertising leaflets show this type of information, often in bright colours (and sometimes intentionally misleading!).

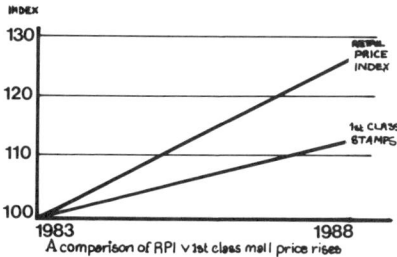

A comparison of RPI v 1st class mail price rises

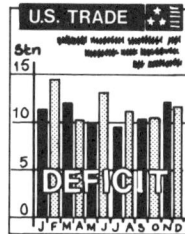

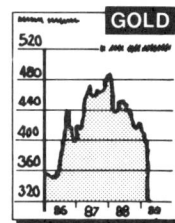

What a bar chart or block graph does is to give an overall picture of a set of numerical information. Your eye can take in a picture of half a dozen columns on a graph, and give a general impression of this information in your mind, but it is much more difficult for you to derive a general impression given by a table of numbers.

Example 1

Draw a bar chart from the information in the table, collected during a survey on colours of cars.

Car colour	white	red	blue	black	brown	green
Number of cars	16	10	11	7	10	6

It does not matter which way we draw the diagram – either way will illustrate the same information. (In mathematics it is convenient to have the number axis vertical.)

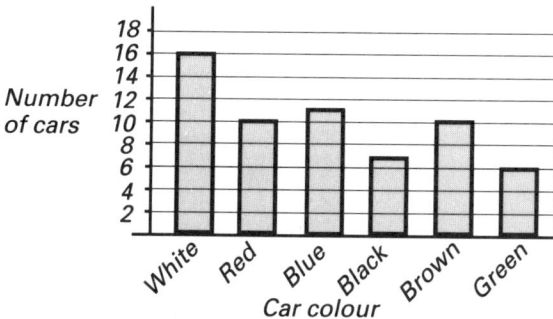

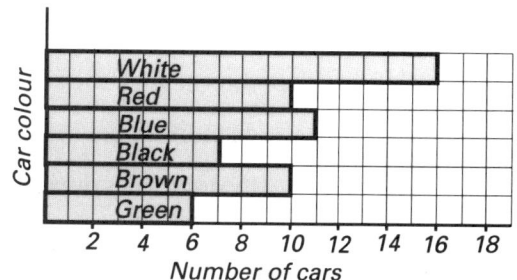

171

In a *bar chart*, it does not matter if there is a gap between the columns.

In a *block graph*, the columns have no gaps; usually there is a scale on the horizontal axis which must not have gaps.

Example 2

As part of her social studies project, Astrid questioned some people as they were going into a polling station to vote. The ages of the group questioned are shown in the bar chart.

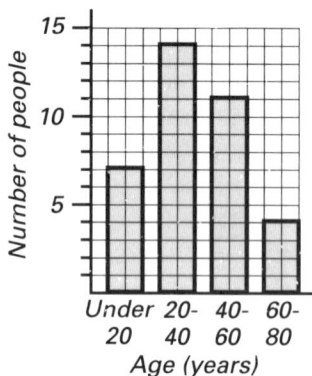

(*a*) How many were aged between 40 and 60?
(*b*) How many were under 40 years old?
(*c*) How many people were questioned by Astrid?
(*d*) What fraction of these people were over 60?

(*a*) From the diagram, 11 people were aged between 40 and 60.
(*b*) The first two bars represent anyone under 40. There were therefore
    $7 + 14 = 21$ people who were under 40 years old.
(*c*) The total number of people questioned was $7 + 14 + 11 + 4 = 36$.
(*d*) The number of people over 60 is 4; the fraction of the total number is
    therefore
    $$\frac{4}{36}\left(= \frac{1}{9}\right).$$

EXERCISE 10.1

1 The graph shows the value of the sales, in thousands of £s, of toys manufactured by a company during the first six months of 1988. For example in January the sales totalled £10 000.

    (*a*) What were the sales in February?

    (*b*) During which month were the sales lowest?

    (*c*) What were the total sales during May and June?

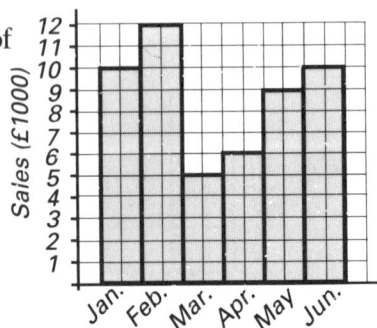

2 A petrol stations sells 4-star, unleaded and super unleaded petrol, as well as diesel. The diagram shows the number of litres of each type sold in one week.

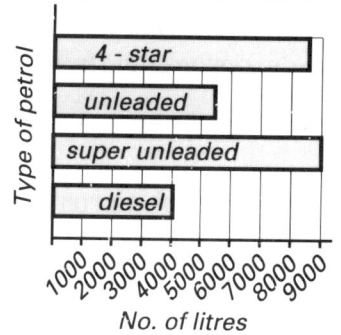

(*a*) How much 4-star petrol was sold?

(*b*) Diesel costs 32p per litre. How much money did the diesel sales bring in to the garage?

(*c*) How many litres of petrol (either 4-star, unleaded or super unleaded were sold during the week?

3 The bar chart shows the different ways pupils in class 4PW came to school today.

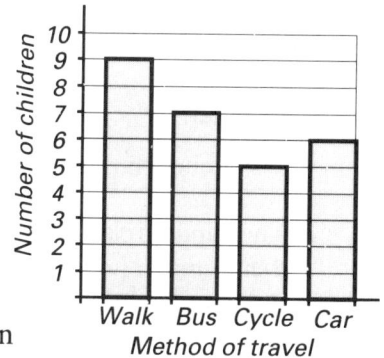

(*a*) How many came by bus?

(*b*) How many either walked or cycled?

(*c*) How many pupils are there in class 4PW?

4 A grocery firm has five vans, A, B, C, D and E. The average weekly distance travelled by each van during last year is shown in the diagram.

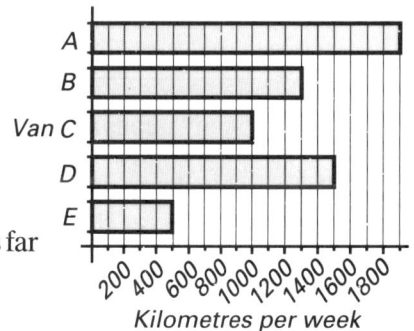

(*a*) How far did van C travel during an average week?

(*b*) One van averaged three times the distance of another. Which van was this?

(*c*) How far would van B expect to travel in four weeks?

(*d*) Work out whether or not vans C and D together went as far as vans A and B.

5 The percentage of votes cast in a by-election is shown in the diagram.

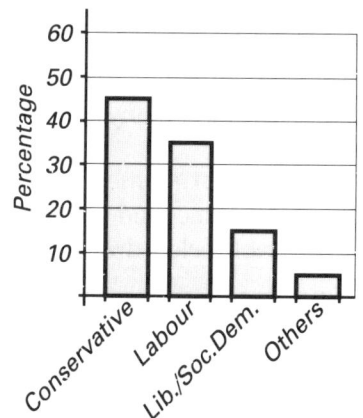

(*a*) What percentage of the votes were for Labour?

(*b*) 'Conservatives win by-election by 20%.' This newspaper headline is incorrect. What was the correct percentage?

(*c*) Work out whether or not the Labour and Lib./Soc. Dem. votes together would have beaten the Conservative votes.

6 Draw a bar graph to show the choice of subject in a fourth year option group from the table below.

Subject	History	Geography	French	Science	Art	Media studies
Number of pupils choosing the subject	28	39	11	44	16	12

7 Yesterday there were four puddings to choose from at lunchtime. The table shows the numbers of each one chosen. Draw a bar graph of this information.

Sweet	Yoghurt	Apple pie	Cheesecake	Prunes
Number chosen	65	40	35	25

8 The pupils in class 4BY were asked where they had been for their summer holiday. From the table which shows their replies, draw a suitable bar graph.

Holiday destination	England	Scotland	Wales	Europe	Other
Number of pupils	8	6	3	8	2

How many pupils are there in class 4BY?

9 The table shows the number of model cars owned by each member of a car club. Draw a suitable bar graph of this information.

Number of cars owned	3	4	5	6	7	8
Number of people	7	2	5	4	4	2

(*a*) How many people are members of the club?

(*b*) How many cars have the members altogether?

10 In an experiment, Sue took one of the dice from a table, threw it 100 times and noted the score each time. From the frequency table below draw a graph of the results.

Score	1	2	3	4	5	6
Frequency	16	22	14	12	17	19

Would you say that the die was 'fair' or 'loaded'?

In Example 1, we could not put a scale on the car colour axis, as you cannot measure car colours numerically (i.e. by metres, or seconds, or grams).

If the horizontal axis does have a scale, as in Example 2, then the diagram is called a **histogram**.

Example 3

A market gardener had a number of Christmas trees for sale. He grouped them by height: between 4 and 5 feet; between 5 and 6 feet; etc. The table shows the number of trees in each group.

Height (feet)	4−5	5−6	6−7	7−8	8−9
Number of trees	8	15	22	18	14

(a) Draw a histogram of this information.
(b) How many trees were over 6 feet tall?
(c) How many trees did he have altogether?

(a) See diagram.
(b) The number of trees over 6 feet tall is

$$22 + 18 + 14 = 54 \text{ trees.}$$

(c) Altogether he had $8 + 15 + 22 + 18 + 14 = 77$ trees.

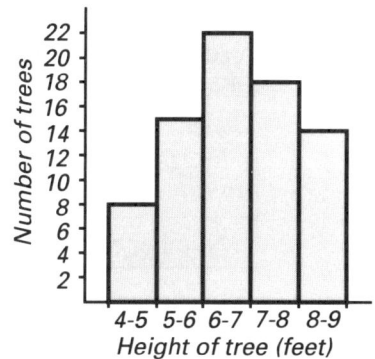

Height of tree (feet)

Example 4

In a survey, pupils were asked how much time, on average, they spent on homework during a week. The results of the survey are shown in the table.

Homework time (hours spent)	0−1	1−2	2−3	3−4	4−5	5−6	6−7	7−8	8−9	9−10
Number of pupils	14	3	7	12	23	5	8	7	4	2

(a) Draw a histogram to show this information.
(b) How many pupils spent less than two hours per week on homework?
(c) How many pupils spent between five and seven hours per week on homework?
(d) How many pupils gave a reply to the survey?

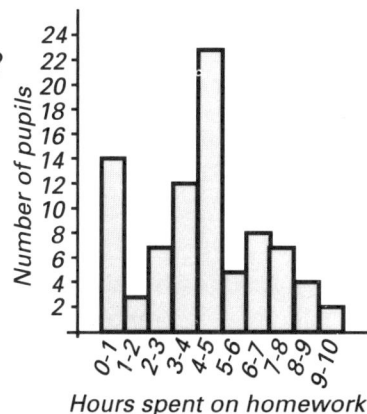

Hours spent on homework

(a) See diagram.
(b) $14 + 3 = 17$ pupils spent less than two hours on homework.
(c) Between five and six hours there were 5 pupils, and between six and seven hours there were 8 pupils. There were therefore 13 pupils who spent between five and seven hours per week on homework.
(d) Total number of pupils replying is

$$14 + 3 + 7 + 12 + 23 + 5 + 8 + 7 + 4 + 2 = 85 \text{ pupils}$$

EXERCISE **10.2**

These questions are about a school 'Sport's Day'. In each case, draw
the bar chart, block graph or histogram, then answer the questions.

1 The table shows the distances reached when parents took part
in a 'throwing the wellie' competition.

Distance thrown	from 0 up to 10 m	from 10 m up to 20 m	from 20 m up to 30 m	from 30 m up to 40 m	from 40 m up to 50 m
Number of throws	8	12	15	7	4

How many parents took part in this competition?

2 A second-year class ran a 'guess the weight of the cake' stall. The
table shows the range of guesses at the weight of the cake.

Weight	from 450 g up to 500 g	from 500 g up to 550 g	from 550 g up to 600 g	from 600 g up to 650 g	from 650 g up to 700 g
Number of guesses	5	10	22	18	11

What do you think is the most likely weight of the cake, to the
nearest 20 g?

3 Some parents joined in a three-legged race. The times of those
pairs who managed to finish are shown below.

Time	from 20 s up to 22 s	from 22 s up to 24 s	from 24 s up to 26 s	from 26 s up to 28 s	from 28 s up to 30 s	from 30 s up to 32 s
Number of teams	2	6	5	9	0	3

If 33 pairs started the race, how many did *not* finish?

4 The times taken by competitors in the 100 m sprint are shown in
the table.

Time taken	under 12 s	from 12 s to 13 s	from 13 s to 14 s	from 14 s to 15 s	from 15 s to 16 s	from 16 s to 17 s	from 17 s to 18 s
Number of competitors	0	4	10	17	8	12	20

(*a*) How many competitors were there in the race?
(*b*) To reach the semi-final you had to finish in under 14 seconds.
   How many competitors reached the semi-final?

5 One third-year group organised a 'Hot Dog' stall. The table shows how many hot dogs were sold during the day.

Time of day	from 1430 up to 1500	from 1500 up to 1530	from 1530 up to 1600	from 1600 up to 1630	from 1630 up to 1700
Number sold	10	55	70	35	10

(*a*) How many hot dogs were sold altogether?
(*b*) The profit on each hot dog was 10p. How much profit did this third-year group make during the day?

6 There were 25 competitors for throwing the discus. The distances thrown are shown below.

Distance thrown	from 10 m up to 14 m	from 14 m up to 18 m	from 18 m up to 22 m	from 22 m up to 26 m	from 26 m up to 30 m	from 30 m up to 34 m	from 34 m up to 38 m
Number of throws	3	12	22	12	18	6	2

Each competitor had the same number of throws, and there were no 'foul throws'. How many throws did each competitor have?

7 The times guessed in the 'Stop the clock' competition run by class 2CF are shown in the table.

Time	from 12.00 up to 2.00	from 2.00 up to 4.00	from 4.00 up to 6.00	from 6.00 up to 8.00	from 8.00 up to 10.00	from 10.00 up to 12.00
Number of guesses	10	14	15	20	12	21

How many people altogether had a guess at the time?

8 In the 'putting the shot' event, the thirty competitors were grouped according to weight. The table shows the distribution of the weights of those competitors who weighed between 45 kg and 70 kg.

Weight	from 45 kg up to 50 kg	from 50 kg up to 55 kg	from 55 kg up to 60 kg	from 60 kg up to 65 kg	from 65 kg up to 70 kg
Number of competitors	4	7	5	8	2

If the lightest competitor weighed 46 kg, how many weighed 70 kg or more?

9 There were 36 starters in the mile race. The times of the finishers are shown in the table.

Time (minutes)	from 4.5 up to 5	from 5 up to 5.5	from 5.5 up to 6	from 6 up to 6.5	from 6.5 up to 7
Number of finishers	6	5	8	10	3

How many did not finish the race?

10 On the sponsored walk around the sport's field, the amount of money collected by those pupils taking part is shown in the table.

Amount collected	from £2 to £3.99	from £4 to £5.99	from £6 to £7.99	from £8 to £9.99	from £10 to £11.99
Number of pupils	20	60	85	45	30

(*a*) How many pupils collected £6 or more?
(*b*) How many pupils took part in the walk?
(*c*) How would you try to estimate the total amount of money raised by these pupils on the sponsored walk?

Line graphs

Example 5

The table shows a patient's temperature taken every six hours, starting at 7 a.m. on Thursday morning.

Time	7 a.m.	1 p.m.	7 p.m.	1 a.m.	7 a.m.	1 p.m.	7 p.m.	1 a.m.	7 a.m.
Temp. (°C)	37.1	39.5	40.2	39.7	39.5	39.5	38.0	37.3	37.1

(*a*) Draw a line graph to represent the patient's temperature.
(*b*) What was her temperature at 7 a.m. on Friday?
(*c*) Estimate at what time on Friday her temperature was 39°C.
(*d*) Estimate for how long her temperature was above 39.5°C.

(*a*) See diagram.
(*b*) Her temperature at 7 a.m. on Friday was 39.5°C.
(*c*) Her temperature was 39°C at about 3 p.m. on Friday.
(*d*) Her temperature was over 39.5°C from 1 p.m. on Thursday until 7 a.m. on Friday, which is 18 hours.

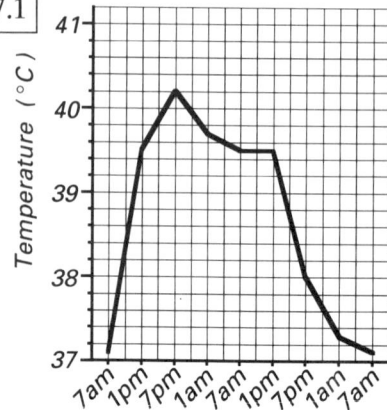

In a line graph which shows something gradually changing, you can estimate some intermediate values, as we did in part (*c*) above.

You must be careful, however, because in some line graphs the lines themselves have no meaning; they are there to show the change from one point to the next.

Example 6

The pupils in a class were asked how many brothers or sisters they each had. The table shows their replies.

Number of brothers or sisters	0	1	2	3	4	5
Number of pupils	7	5	9	4	2	1

(*a*) Draw a line graph of this information.
(*b*) How many pupils have two or more brothers or sisters?
(*c*) How many of the pupils are 'an only child'?
(*d*) How many pupils are there in the class?

(*a*)

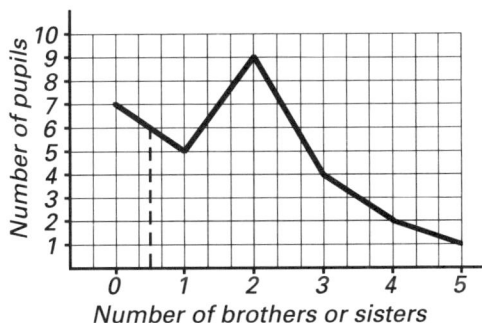

(*b*) There are $9 + 4 + 2 + 1 = 16$ pupils with two or more brothers or sisters.
(*c*) There are 7 'only' children.
(*d*) There are $7 + 5 + 9 + 4 + 2 + 1 = 28$ pupils in the class.

In the line graph the lines joining the points have no meaning; however, they enable us to see more clearly the changes from one number to the next. If you look at the dotted line, it is meaningless to say that six pupils had 'half a brother or sister' – they may have a half-sister, but that is something entirely different!

Take care, therefore, when reading line graphs.

EXERCISE 10.3

1 The depth of water at a pier head is read every hour. The diagram shows the readings.

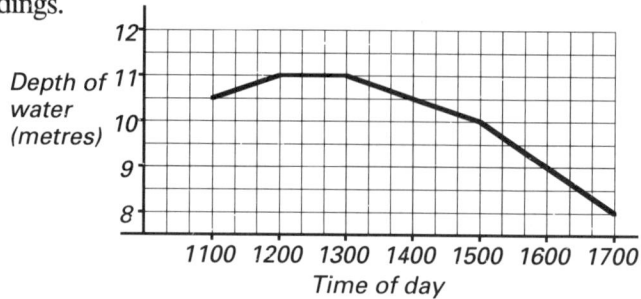

(*a*) What is the depth at 1400 hours?

(*b*) At what time is the depth 9½ metres?

(*c*) At what time is 'high tide'?

2 A baby chimpanzee was weighed every week for six weeks. From the graph estimate

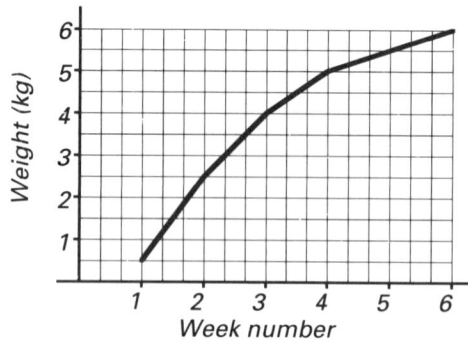

(*a*) its weight after three weeks,

(*b*) its *increase* in weight from week 3 to week 6.

3 The graph shows the profits of a printing firm over a period of five years.

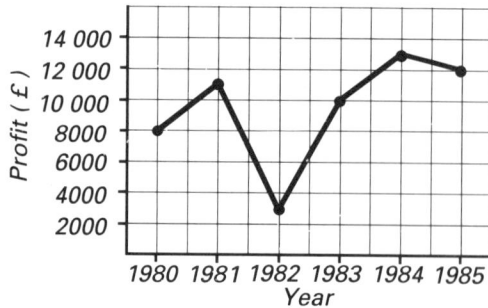

(*a*) What was the profit in the best year?

(*b*) Which year showed £10 000 profit?

(*c*) By how much did the profit fall from 1981 to 1982?

4 As part of a humanities project, the rainfall during the six-week summer break was measured weekly.

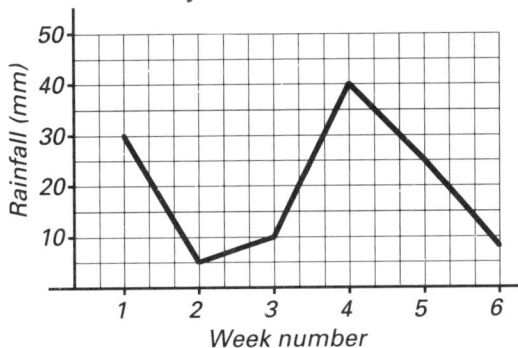

(*a*) What was the rainfall during week 3?

(*b*) Which was the driest week?

(*c*) How much rain fell during the six weeks?

(*d*) What was the average weekly rainfall, correct to the nearest millimetre?

5 During the first week of the summer break, another pupil measured the outdoor shade temperature at noon each day. Use the diagram to estimate

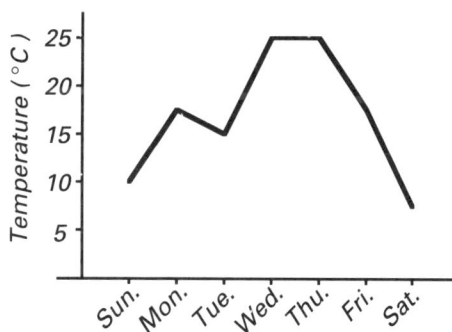

(*a*) the noon temperature on Tuesday,

(*b*) the hottest day (or days) at noon,

(*c*) the average noon temperature for the week (approximately).

6 Draw a line graph of the temperature of a patient in hospital, as shown in the table.

Time	0900	1500	2100	0300	0900	1500	2100	0300
Day	Mon.	Mon.	Mon.	Tue.	Tue.	Tue.	Tue.	Wed.
Temp. (°C)	37.2	38.6	37.7	37.3	38.0	37.8	37.2	37.0

7 The number of first-year pupils entering a school each year from 1980 to 1986 is shown in the table. Plot the points and draw a line graph.

Year	1980	1981	1982	1983	1984	1985	1986
Intake	340	300	310	250	270	230	230

8 On the diagram you have drawn for question 7, draw, in a contrasting way (dotted or coloured), a line graph of sixth-form numbers, as shown in the table.

Year	1980	1981	1982	1983	1984	1985	1986
Sixth-formers	120	130	85	90	75	100	140

9 On a fishing holiday, Angus wrote down how many fish he caught each day. Draw a line graph from the information in the table.

Day	Mon.	Tues.	Wed.	Thu.	Fri.	Sat.
Number of fish	6	10	0	8	8	3

How many fish did Angus catch during the week?

10 The table shows the winning times in the 50 m freestyle race at our school's annual swimming gala, from 1980 to 1986. Draw a line graph of these times.

Year	1980	1981	1982	1983	1984	1985	1986
Winning time (s)	36.4	38.1	34.5	35.3	35.3	33.0	33.9

In which year was the winning time of the previous year beaten by more than three seconds?

Pie charts

Example 7

An electrical department in a hypermarket sold the following number of light bulbs in a week.

Type of bulb	40 W	60 W	100 W	150 W
Number sold	20	40	30	10

Show this information (*a*) in a bar chart (*b*) in a pie chart.

(*a*) A bar chart could look like this:

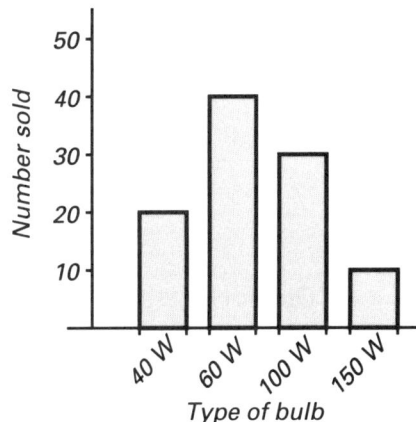

(*b*) In the diagram below, the 'pie' is divided into four sectors. The size of each sector represents the number of the type of bulb in that sector.

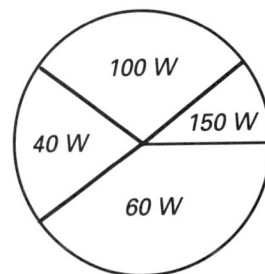

With a small number of different items, a pie chart can be useful in giving a quick visual comparison.

Example 8

The pie chart shows the percentage of different sizes of nails that I have in a box.
(*a*) What percentage of nails have a length of 10 cm?
(*b*) What percentage of nails are of length 5 cm?
(*c*) What fraction is this?
(*d*) There are 200 nails in my box. How many of them have a length of 15 cm?

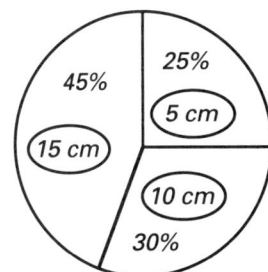

(*a*) 30% of the nails are of length 10 cm.
(*b*) 25% are of length 5 cm.
(*c*) $25\% = \frac{25}{100} = \frac{1}{4}$.
(*d*) 45% of 200 = 90 nails are of length 15 cm.

EXERCISE 10.4

1 The percentage of small, medium and large jars of coffee sold during one Thursday at a supermarket is shown in the pie chart. If 100 jars were sold altogether, how many were small ones?

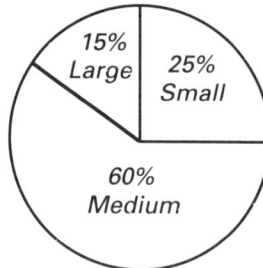

2 The pie chart shows the percentage of grades A, B or C awarded in a science coursework assignment.

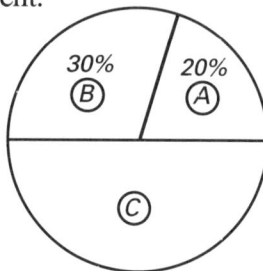

(*a*) What percentage of C grades were awarded?

(*b*) Fifty pupils achieved an A, B or C. How many of these were A's?

3 There are four possible answers to a multi-choice science question. When fourth-year pupils attempted the question, 10% gave answer A, 10% gave answer B, and 60% gave answer C.

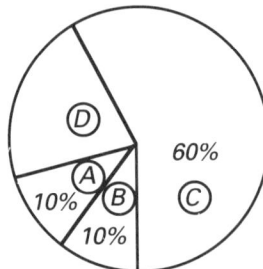

(*a*) If every fourth-year pupil gave an answer, what percentage gave answer D?

(*b*) There are 300 pupils in the fourth year. How many gave answer C?

4 The pie chart shows the percentages (by volume) of cement, lime and sand that should be mixed in order to form a mortar suitable for bricklaying.

(*a*) How much lime should I mix with 750 cm³ of sand?

(*b*) If instead I require 2000 cm³ of mortar, how much cement will I need?

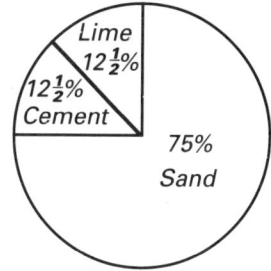

5 In a GCSE course, 40% of the marks are available on paper 1. The remaining marks are allocated equally to paper 2, the practical test and the project.

(*a*) What percentage of marks is for paper 2?

(*b*) What percentage is for coursework (i.e. practical + project)?

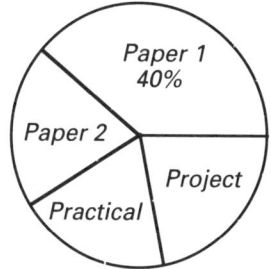

6 The table shows the percentage of men, women and children on a Channel ferry. Draw a pie chart of the information. (Make the radius of the circle about 4 cm.)

Passenger	Men	Women	Children
Percentage	40	30	30

7 Ann and Asif sold raffle tickets at the school disco. There were four colours on sale: blue, yellow, white and green. The table shows the percentage of each colour sold. Draw a pie chart, labelling each sector correctly.

Colour	Blue	Yellow	White	Green
Percentage sold	15	35	20	30

8 The 50 pupils in classes 3JP and 3JT are divided into three groups for humanities. There are 15 pupils in group A, 15 in group B and 20 in group C. Copy and complete the table, and draw a pie chart.

Group	A	B	C
Number of pupils	15	15	20
Percentage	30%		

9 The number of pupils sitting each of the four mathematics papers in a GCSE examination is given in the table. Copy and complete the table, and draw a pie chart. Remember to label the sectors.

Paper	1	2	3	4
Number sitting paper	28	80	72	20
Percentage	14%			

10 For each £1 of income, a council spends 40p on education, 25p on housing, 15p on welfare and the remainder on other services. Make up a table, and draw a pie chart, labelling the sectors.

Pictograms

Houses needed

Houses built

(= 100 houses)

'750 houses needed – only 450 built'
The use of a sketch of an object (a house) to represent a number of houses (100 in this case) is an example of the use of a **pictogram** to give a visual image of a set of data.

Instead of using the same size of house seven-and-a-half times, some pictograms use different sizes of house to convey the same information. This method has to be carefully interpreted, otherwise a false impression can be created. This is sometimes the intention of the person presenting the diagrams!

In diagram A the *lengths* of the houses shown are in the correct proportion, but in diagram B it is the *areas* of the figures which are in the correct proportion.

A

B

Example 9

Answer these questions from the diagram.
(*a*) How many planes does airline A have?
(*b*) How many more planes does airline C have than airline B?
(*c*) Two airlines combine, to give a total of 1200 planes. Which two?
(*d*) How many planes do the five airlines have altogether?

= 100 aeroplanes

(*a*) Airline A has 450 planes.
(*b*) Airline C has 800 and airline B has 650, so C has 150 more than B.
(*c*) The only two airlines which give a total of 1200 are C and E.
(*d*) Altogether there are 450 + 650 + 800 + 500 + 400 = 2800 planes.

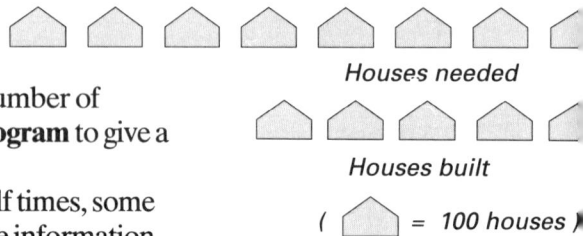

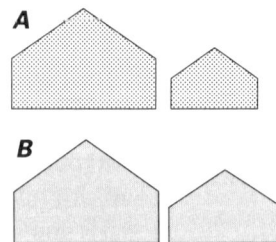

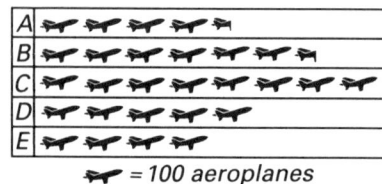

EXERCISE 10.5

1 The diagram represents the number of cars produced by a motor manufacturer during the first three weeks in April.

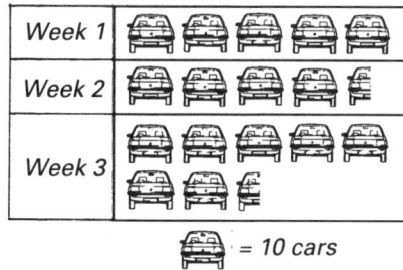

= 10 cars

(*a*) How many cars were produced during week 2?

(*b*) How many cars were produced in these three weeks altogether?

2 In a small town there are two bakeries, Brown's and White's. The diagram shows the number of large loaves sold during one week.

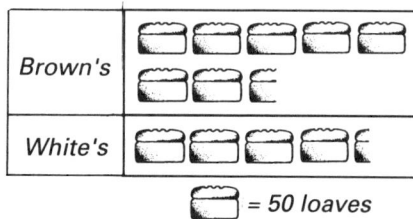

= 50 loaves

(*a*) How many loaves did Brown's sell?

(*b*) How many loaves were sold altogether?

3 The increase in sales of books from a bookshop over four years is shown in the diagram.

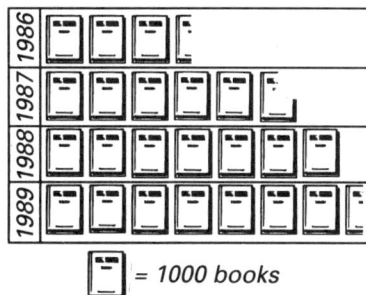

= 1000 books

(*a*) How many books were sold in 1986?

(*b*) How many *more* books were sold in 1988 than in 1987?

(*c*) Estimate the 1990 sales.

4 Grey's, a grocery firm, compared its sales of low-fat spread with those of two rival firms. The diagram shows the sales, in kilograms.

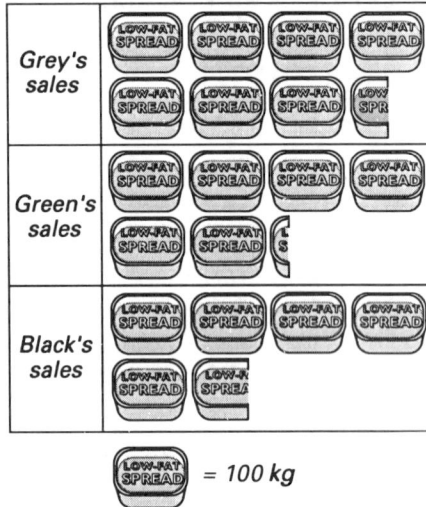

= 100 kg

(*a*) Whose sales were greatest?

(*b*) What were Green's sales?

(*c*) What were Black's sales?

5 Caroline classifies her cassettes into Pop, Sixties, Heavy Metal and others. From the diagram estimate

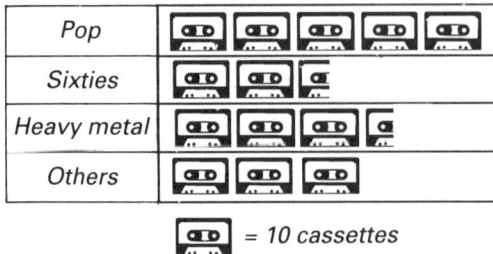

= 10 cassettes

(*a*) how many Sixties cassettes she has,

(*b*) how many cassettes she has altogether.

Frequency distributions

From an 11 – 16 comprehensive school of five year groups, eight pupils were selected at random to take part in a survey on healthy eating. The pupils were:

Jane Barker (5CX)	Norman Henderson (1FD)
Badar Maneer (3HW)	Judith Robinson (5CX)
James McMahon (1FT)	Andrew Harvey (2SY)
Katriona Harland (3MM)	Melanie Smith (3HT)

If we wish to know how many pupils were chosen from each year group, a table would make it easier to see.

Year group	1st	2nd	3rd	4th	5th
Number of pupils	2	1	3	0	2

This is an example of a **frequency distribution** or **frequency table**, because it shows how *frequently* the pupils in each year group are *distributed* over the range from the 1st to the 5th year.

We can draw a bar chart of this information so that we have a picture of the distribution of pupils over the range.

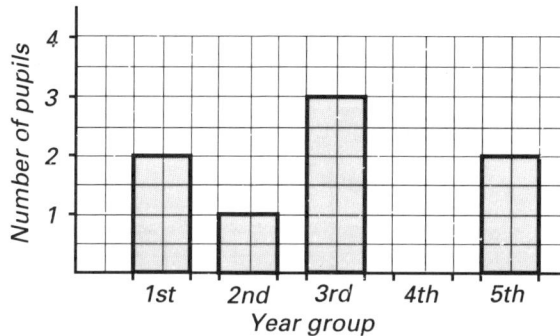

If 50 pupils had been selected, the frequency table may possibly have looked like this:

Year group	1st	2nd	3rd	4th	5th
Number of pupils (or frequency)	11	13	8	11	7

The frequency distribution graph would have looked like this.

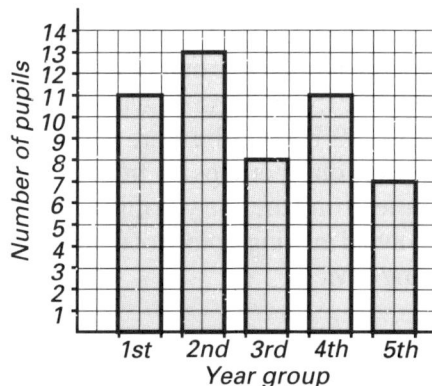

Example 10

As part of a GCSE assessment in art, at least five pieces of work are required. Last year the number of pieces produced by the fifteen pupils in one group were:

8, 5, 5, 6, 8, 7, 10, 5, 7, 7, 8, 6, 5, 8, 8.

(a) Put these figures into a frequency table.
(b) Draw a frequency diagram.

(a) The table looks like this.

Number of pieces of work	5	6	7	8	9	10
Number of pupils (frequency)	4	2	3	5	0	1

(b) This is the frequency diagram.

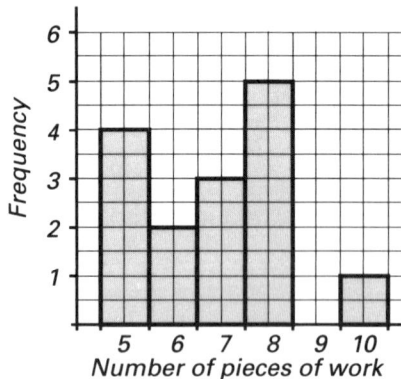

Tally marks

Before trying some frequency tables, there is a useful technique which you should adopt when you have to deal with large numbers of data.

In the above example there were just 15 pupils, so drawing up the table was quite easy. With more pupils, say 50, it would be more difficult to check that you had accounted for each one; perhaps the frequencies add up to 49, and it takes you a long time to find the one you missed!

The use of **tally marks** helps to avoid missing a piece of data, and is both a useful check and time-saving device.

Example 11

The marks scored by 30 pupils in a science test are given below.
(The maximum mark was 10.)

3	6	6	4	9	10
6	7	10	8	4	5
8	0	3	4	8	7
8	10	6	3	4	8
1	5	8	5	9	8

(*a*) Using tally marks, make up a frequency table of marks scored.
(*b*) Draw the frequency diagram.
(*c*) If you scored less than 5 you had to repeat the test later.
How many pupils needed to repeat the test?
(*d*) To gain a 'credit' you need to score 8 or more. How many pupils obtained a credit on this test?

(*a*) It is usual when using tally marks to write the scores down the page instead of across, and to make three columns:

Score	Tally marks	Frequency
0		
1		
2		
3		
4		
5		
6		
7		
8		
9		
10		

Now go through the numbers in the table systematically, making a vertical mark (tally mark) in the correct place in the middle column of the table for each number as you come to it. (You can put a pencil line through each number as you come to it.)

The tally table overleaf shows the tally marks for the first column, with marks shown for the numbers 3, 6, 8, 8 and 1.

Score	Tally marks	Frequency
0		
1	¦	
2		
3	│	
4		
5		
6	│	
7		
8	‖	
9		
10		

The completed table looks like this.

Score	Tally marks	Frequency
0	│	1
1	│	1
2		0
3	‖|	3
4	‖‖	4
5	‖│	3
6	‖‖	4
7	‖	2
8	ℍℾ ‖	7
9	‖	2
10	‖|	3
Total		**30**

Look carefully at the '8' score. When you get to four tally marks, make the fifth one go across, like a farm gate. We then count this as a block of five.

Although in this example the use of the block of five makes little difference, it can be of great help when you have many more figures to tally. You can then count them in fives, rather than as single marks.

Also, it is worthwhile, as a check, to add the final frequencies, in order to ensure that the total of the tally marks is correct (30, in this example).

(b) You can now draw the frequency diagram.

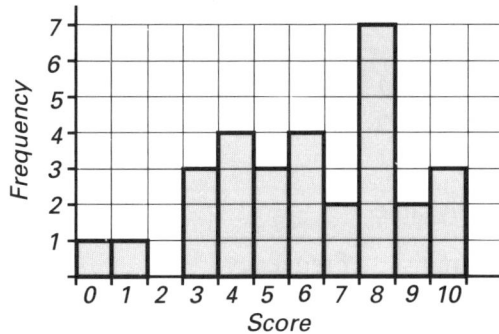

(c) There are $1+1+0+3+4=9$ pupils who will have to repeat the test.

(d) $7+2+3=12$ pupils gained a credit.

EXERCISE **10.6**

1 The number of matches in 40 boxes is shown below.

42	38	45	41	42	45	37	45	39	41
42	40	40	41	38	44	43	41	40	41
43	41	41	44	41	40	43	44	42	38
40	39	45	40	41	42	43	41	42	40

(a) Copy and complete the tally table.

No. of matches	Tally marks	Frequency
36		
37		
38		
39		
40	l	
41		
42	ll	
43	l	
44		
45		
Total		

(b) Draw a bar chart.

(c) How many boxes contained 42 matches?

(d) How many boxes contained *less* than 40 matches?

2   The table shows the number of goals scored by our hockey team in 30 matches played this season.

```
4   2   3   3   7   0   1   0   2   2
3   1   2   1   4   6   2   0   0   3
1   2   0   3   4   2   6   4   7   0
```

(*a*) Copy and complete the tally table.

Score	Tally marks	Frequency
0		
1		
2		
3		
4		
5		
6		
7		
Total		

(*b*) In how many matches did we score three goals?

(*c*) How many goals did we score altogether?

(*d*) How many matches can you be *sure* that we did *not* win?

3   The number of tomatoes picked from each of twenty plants is shown below.

```
23   28   30   27   25   24   25   27   23   28
27   23   22   24   27   28   26   26   27   26
```

(*a*) Complete the tally table.

No. of tomatoes	Tally marks	Frequency
22		
23		
24		
25		
26		
27		
28		
29		
30		
Total		

(*b*) Draw a bar chart.

(*c*) How many plants had more than 27 tomatoes picked from them?

4 The 36 students beginning a college course were asked to write down the numbers of GCSE certificates they had obtained. The figures are shown below.

8	6	7	6	9	8	5	7	6
8	7	10	5	6	7	7	5	7
9	8	6	9	7	8	8	6	7
7	10	9	7	8	7	5	6	6

(*a*) Copy and complete the tally table.

Number of GCSE certificates	Tally marks	Frequency
5		
6		
7		
8		
9		
10		
Total		

(*b*) Draw a bar chart.

(*c*) How many students had at least eight GCSE certificates?

5 At a village field day, the local garage offered to give a new car to anyone who threw five 6's with one throw of five dice. The number of 6's thrown in each of the first 40 attempts is shown below.

1	0	0	1	2	1	1	0	0	1
0	1	3	1	0	0	1	2	1	0
0	1	0	3	0	2	1	2	0	1
1	0	0	1	4	0	1	1	0	2

(*a*) Make up a tally table like the one in question 3, with the number of 6's thrown in the first column.

(*b*) Draw a bar chart.

(*c*) How many times were no 6's thrown?

(*d*) On average, how many throws do you think would be needed in order for someone to win the car?

In the questions and examples so far in this section we have considered whole number possibilities only, such as number of pieces of work completed, marks scored in a test, number of pets, etc. These are called **discrete** (or countable) values.

If the information is obtained by measuring rather than counting, then the values are **continuous** (or measurable).

Example 12

The heights of twenty young tomato plants in a garden centre were measured in millimetres. The heights were:

93　86　54　76　90　58　63　71　80　98
84　83　69　76　74　70　85　51　81　78

The frequency table would be:

Height (mm)	51	52	53	54	55	56	57	58	59	60	61	62	63	64	...	98	99
Frequency	1	0	0	1	0	0	0	1	0	0	0	0	1	0	...	1	0

A bar chart of the heights would look like this:

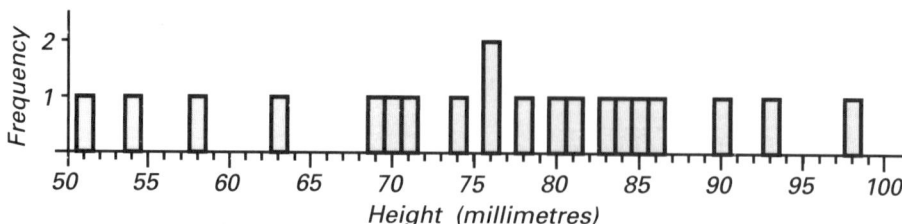

This diagram does not tell us much about the heights of the plants, other than that they vary intermittently from 51 mm to 98 mm. It would be better if we had the heights grouped in blocks of, say, 5 mm. We would then have ten groups of heights.

Height (mm)	50–54	55–59	60–64	65–69	70–74	75–79	80–84	85–89	90–94	95–•
Frequency	2	1	1	1	3	3	4	2	2	1

The frequency diagram from this table is rather clearer than the previous one; at least we can see that there are more heights at the higher end of the range.

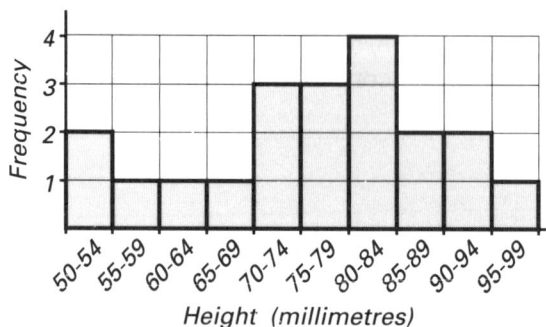

Height (millimetres)

Perhaps it would be better to take groups of ten.

Height (mm)	50–59	60–69	70–79	80–89	90–99
Frequency	3	2	6	6	3

This frequency diagram is better still. It gives a fairly clear picture of the spread of heights; over half are fairly evenly spaced between 70 mm and 90 mm, with rather more of the remaining heights under 70 mm than over 90 mm.

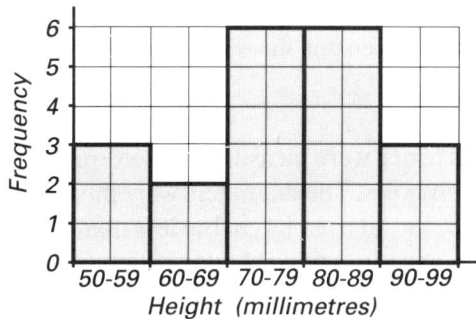

The diagrams drawn so far are still block graphs. If we wish to put a scale along the horizontal axis, how could we do it?

We could not mark the scale with 50, 60, 70, 80, 90 and 100 at the bottom of each bar, because we would not know into which block to put the height of 70 mm, or 80 mm.

There are two possible answers.

(i) Make the divisions between the blocks half-way between the highest value in one block and the lowest in the next. This will mean that the vertical lines are drawn at 59.5 mm, 69.5 mm, 79.5 mm and 89.5 mm, as in the diagram.

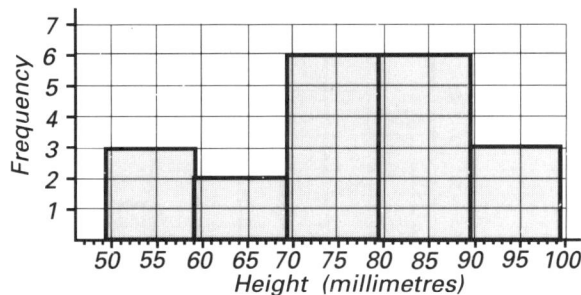

(ii) Draw the vertical lines at 50, 60, etc., then *include* the lowest value and *exclude* the highest value. Thus 70 mm would be in the 70–80 mm block, but 80 mm would be in the 80–90 mm block, as shown in the diagram overleaf.

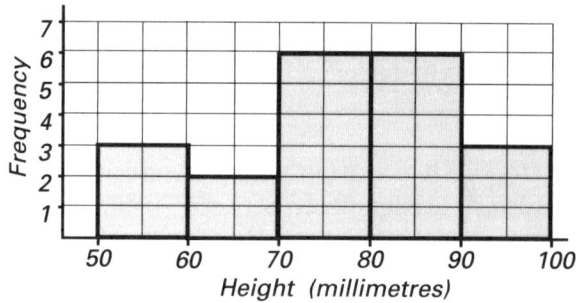

Although there are mathematical differences between these methods, there is little practical difference when you have large numbers of values.

Example 13

The diameters of 60 trees in a forest were measured to determine which would be suitable for telegraph poles. The diameters were measured as 'at least 17 cm but less than 18 cm', 'at least 18 cm but less than 19 cm', etc.

The table shows the frequencies, in groups of one centimetre.

Diameter (cm)	from 17 to 18	from 18 to 19	from 19 to 20	from 20 to 21	from 21 to 22	from 22 to 23	from 23 to 24
Frequency	7	10	11	13	10	6	3

(*a*) Draw a frequency diagram.

(*b*) Trees with diameters between 20 cm and 23 cm are suitable for telegraph poles. How many trees are suitable?

(*a*) The vertical lines of the blocks can be drawn at 17 cm, 18 cm, etc.

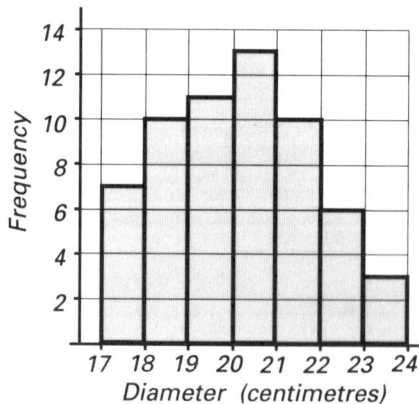

(*b*) There were $13 + 10 + 6 = 29$ trees suitable for telegraph poles.

EXERCISE **10.7**

1 A coal lorry carried 30 sacks; each one was supposed to contain
50 kg of coal. The actual weights of the sacks, to the nearest
kilogram, were:

```
52  54  51  52  50  53  48  49  52  54
56  50  53  51  51  52  53  48  51  53
50  53  55  52  48  51  51  52  50  54
```

(*a*) Copy the frequency table and complete it with tally marks.

Weight (kg)	Mid-value	Tally marks	Frequency
47.5 – 48.5	48		3
48.5 – 49.5	49		
49.5 – 50.5	50		
50.5 – 51.5	51		
51.5 – 52.5	52		
52.5 – 53.5	53		
53.5 – 54.5	54		
54.5 – 55.5	55		
55.5 – 56.5	56		
		Total	

(*b*) Draw a frequency diagram (take care with the horizontal
scale).

(*c*) How many sacks weighed less than 49.5 kg?

2 Frank, as part of his social studies assignment, asked the ages of
40 people going into a supermarket. Their replies are given
below.

```
18  19  41  53  26  33  25  18  17  23
26  45  63  63  46  37  28  29  27  16
29  40  45  35  38  28  20  42  39  38
19  21  24  40  26  37  47  42  37  29
```

(*a*) Copy and complete the tally table.

Age range	Tally marks	Frequency
10–19		
20–29		
30–39		
40–49		
50–59		
60–69		
	Total	

(*b*) Draw a frequency diagram.

(*c*) How many people were in their fifties?

(*d*) How many were under 30 years old?

3 As these 40 people emerged from the supermarket, Frank asked them how much they had spent. Their answers are given below.

£15.76	£52.63	£24.53	£8.92	£16.02
£10.70	£33.57	£19.46	£12.72	£21.80
£40.52	£28.88	£36.63	£13.29	£2.99
£18.49	£38.62	£42.73	£31.74	£11.24
£14.80	£20.66	£17.04	£36.29	£10.63
£18.20	£6.46	£16.33	£29.05	£25.81
£58.31	£22.77	£45.62	£38.01	£10.42
£18.62	£33.38	£28.01	£5.74	£44.77

(*a*) Copy and complete the tally table.

Amount spent	Tally marks	Frequency
£0–£9.99		
£10.00–£19.99		
£20.00–£29.99		
,, – ,,		
,, – ,,		

(*b*) Draw a frequency diagram of this information.

(*c*) How many people spent over £50?

(*d*) How many spent under £20?

(*e*) For these 40 customers, what would you say the average bill was, approximately?

4 There were 152 finalists in a national crossword competition. The times, in minutes, taken by the 60 competitors who managed to finish the crossword, are shown below. For example, a time of 1 minute means any time during the first 60 seconds, and a time of 2 minutes means any time from 1 minute after the starting time until 2 minutes after the starting time.

```
28  23  26  17  21  18  24  26  21  14
27  17  21  25  22  16  24  26  19  20
17  18  23  19  20  25  27  26  28  23
24  23  19  20  23  28  22  22  17  22
18  19  27  16  24  25  23  23  23  22
26  22  28  21  19  18  24  23  26  25
```

(*a*) Make up a tally table.

(*b*) Draw a frequency diagram from this table. (You will need to be careful with the horizontal scale.)

(*c*) How long did it take the winner to complete the crossword?

(*d*) If you completed the crossword in under 20 minutes you received a prize. How many competitors received a prize?

5 The number of league goals scored by each of the 44 teams in Divisions 1 and 2 of the Football League during the 1987 – 8 soccer season is given below.

```
87  71  67  53  48  58  58  55  57  46  52
49  38  40  35  40  38  50  36  27  44  72
68  63  74  68  86  61  61  80  72  50  73
62  61  54  65  56  42  41  50  45  44  41
```

(*a*) Copy and complete the tally table.

Number of goals	Tally mark	Frequency
0-9		
10–19		
20–29		
30–39		
,, – ,,		
,, – ,,		

(*b*) Draw a frequency diagram to show the number of goals scored.

(*c*) How many teams scored at least 70 goals?

(*d*) How many teams scored less than 40 goals?

Mode and median

There are eleven junior players in a badminton club. The number of games played by each one during a season is given below.

Amanda	9	Sally	5	Jim	9
Donna	8	Cliff	7	Suzy	12
Harry	9	Rob	10	Julia	7
Mary	10	Lucy	9		

What is the average number of games played by a junior player?
It would be clearer if we put the information in a frequency table.

Games played	5	6	7	8	9	10	11	12
Number of players	1	0	2	1	4	2	0	1

There are three types of 'average' that can be considered:

(i)   the **mode**
(ii) the **median**
(iii) the **mean**.

You may have met these before, but here is a reminder.

(i) The **mode**, or **modal value**, is the 'most popular' value, or the value occurring most often.
   In this example the mode is 9 games, as more players played 9 games than played any other number of games.

(ii) The **median** is the middle value when all the values are arranged in ascending order of magnitude.
   In this example the ordering is: 5, 7, 7, 8, 9, 9, 9, 9, 10, 10, 12. The middle value is the sixth one along, so the median is 9 games.

1st	2nd	3rd	4th	5th	6th	7th	8th	9th	10th	11th
5	7	7	8	9	9	9	9	10	10	12

Median value

   (So long as there is an *odd* number of values, there will be a middle value. If there is an *even* number of values, then the median is half-way between the two middle values.)

(iii) The **mean** requires a calculation (the mode and median do not, so are easier to find); it is obtained by adding all the values, and then dividing this total by the number of values that we have added up (i.e. by the total frequency of the distribution).

   In this example, the mean value is

$$\frac{5+7+7+8+9+9+9+9+10+10+12}{11} = \frac{95}{11} = 8.636 \ldots \text{ or about 8.6 games.}$$

EXERCISE *10.8*

1 Work out the mode and the median for each of these sets of data.

(*a*) 3, 4, 4, 4, 5, 6, 6, 8, 9

(*b*) 21, 23, 24, 24, 25, 26, 27, 27, 27, 29, 38

(*c*) 5, 8, 4, 6, 8

2 A group of nine pupils received the following marks in a French vocabulary test: 13, 15, 20, 13, 16, 15, 18, 13, 12.

(*a*) Re-write the marks in order (12, 13, ..., 20).

(*b*) State the mode (modal mark).

(*c*) What is the median mark?

3 (*a*) Write down a set of seven numbers in which the mode is 8 and the median is 10.

(*b*) Repeat, this time with six numbers.

4 Twelve pupils in a class made a guess at the length of a chalk line drawn across the school yard by the teacher. Their guesses, in metres, were: 8, 5, 6, 6, 4, 7, 7, 4, 7, 5, 6, 7.
Work out the modal length, and the median length of the guesses.

5 The number of points scored by each of the nine players in our under-15 basketball team so far this season is:

12, 11, 13, 26, 15, 6, 11, 12, 11.

Find (*a*) the median number of points scored (*b*) the modal number of points scored.

Investigation

Find the mean value of each set of numbers in Exercise 10.8, questions 1, 2, 4 and 5.

Quartiles

When a set of data is arranged in numerical order, as we have just seen, the central value is called the median. If we just consider the lower half of the arranged data, its middle value is called the **lower quartile**. Similarly, the middle value of the upper half is called the **upper quartile**.

Looking at the junior badminton players, the data, written in order, is:

  5, 7, 7, 8, 9,   9,   9, 9, 10, 10, 12

The middle value of the lower half is 7, while the middle value for the upper half is 10.

  So the lower quartile is 7 games, and the upper quartile is 10 games.

1st	2nd	3rd	4th	5th	6th	7th	8th	9th	10th	11th
5	7	7	8	9	9	9	9	10	10	12

                Lower quartile   Upper quartile

EXERCISE **10.9**

For each question in Exercise 10.6, find the median value, and the quartile values.

Grouped data – mode, median

What happens if we are given a frequency distribution table of grouped data, and wish to find the mode and the median?

  Look at Example 13, concerning the diameter of trees. As we do not know the individual diameters, we can state only the *group* which occurs more frequently than any other.

  The *modal group*, therefore, is the 20 cm – 21 cm group, as this occurs most often (13 times).

  For the median, we need to find the diameter of the middle tree; in this example there are two 'middle' trees, the 30th and 31st when put in order.

  Now the smallest 7 trees are between 17 cm and 18 cm, the next smallest 10 are up to 19 cm, and the next 11 are up to 20 cm. So the first 28 trees in the table have diameters less than 20 cm. The 30th and 31st trees must be between 20 cm and 21 cm, because the next 13 are in this range.

If we assume that the diameters are evenly distributed from 20 cm to 21 cm, the median diameter is only a little over 20 cm. In fact it is $\frac{2.5}{13}$ cm over 20 cm (why 2.5?). So the *median* diameter is approximately 20.2 cm $(\frac{2.5}{13} = 0.19\ldots)$.

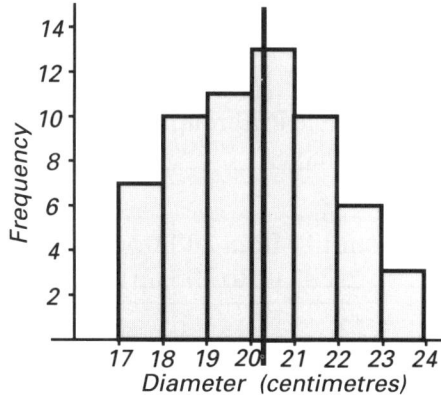

EXERCISE **10.10**

In each of these questions state the modal group, and estimate the median value.

1 The table gives the distribution of heights of leek plants, in centimetres.

Height (cm)	from 10.0 up to 10.5	from 10.5 up to 11.0	from 11.0 up to 11.5	from 11.5 up to 12.0	from 12.0 up to 12.5
Frequency	8	22	40	17	12

2 Boxes of raspberries were weighed, in kilograms. The results are shown in the table.

Weight (kg)	from 1.1 up to 1.2	from 1.2 up to 1.3	from 1.3 up to 1.4	from 1.4 up to 1.5	from 1.5 up to 1.6	from 1.6 up to 1.7
Frequency	3	5	9	10	17	9

3 A group of weightwatchers were weighed. The table shows their weights.

Weight (kg)	from 65 up to 70	from 70 up to 75	from 75 up to 80	from 80 up to 85	from 85 up to 90	from 90 up to 95
Frequency	7	12	20	7	2	1

4 In the school athletics competition, the lengths of the discus throws were measured, as shown in the table.

Length (m)	from 20 up to 25	from 25 up to 30	from 30 up to 35	from 35 up to 40	from 40 up to 45	from 45 up to 50
Frequency	11	16	12	14	5	1

5 The time, in seconds, for each child in a group to complete a puzzle is given in the frequency table.

Time (s)	from 0 up to 10	from 10 up to 20	from 20 up to 30	from 30 up to 40	from 40 up to 50	from 50 up to 60
Frequency	3	12	15	19	7	3

Scatter graphs

There are occasions when there is some connection between two different measurable characteristics; for example, height and weight. We would expect taller people on the whole to be heavier than shorter people, although this is clearly not always the case.

We can obtain a picture of the connection, or relationship, if there is one, by making a **scatter graph** of the information, and seeing if there appears to be any trend.

On a scatter graph each cross indicates the two attributes (say height and weight) of one person.

You should indicate one characteristic along the horizontal axis and the other on the vertical axis.

Example 14

The heights and weights of the eleven junior badminton players are given below. Draw a scatter graph of the information.

	Amanda	Sally	Jim	Donna	Cliff	Suzy	Harry	Rob	Julia	Mary	Lucy
Height (cm)	152	144	158	168	150	139	148	151	130	145	161
Weight (kg)	44	43	48	49	45	40	43	42	39	41	46

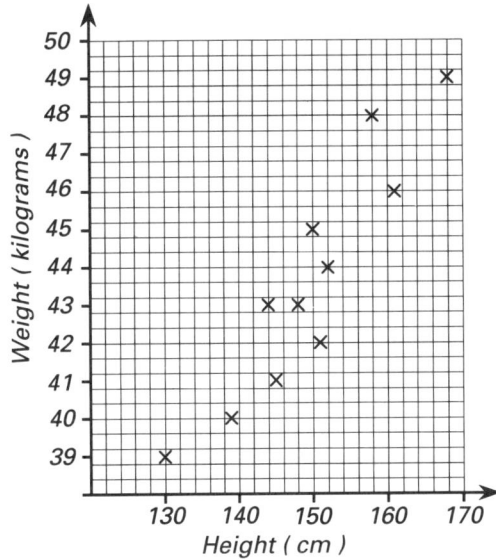

From the graph it appears that the taller you are the heavier you are likelier to be – as you move along the height axis, you tend to move up the weight axis.

Line of best fit

On the scatter graph in Example 14, we could attempt to draw a line which represents the trend. This line, which should have as many crosses above it as it has below, and makes the total of the distances of the crosses from the line as small as possible (ignoring any 'rogue' points wildly out), is called the **line of best fit**.

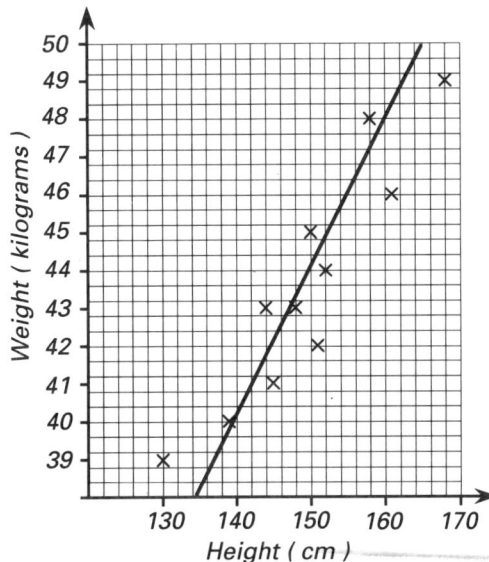

EXERCISE **10.11**

1 On Sport's Day there were nine competitors in both the 100 m race and in the long jump. Draw a scatter graph, and line of best fit, from the figures given in the table.

100 m time (seconds)	12.3	12.3	12.5	12.6	12.9	13.4	13.7	13.7	14.3
Long jump (metres)	6.0	5.1	5.4	4.9	5.0	4.6	4.6	4.1	3.8

2 The marks scored by 12 pupils in paper 1 and paper 2 of a geography examination are given below.

Paper 1	35	56	87	43	52	94	23	35	48	56	66	49
Paper 2	43	69	72	62	51	88	24	40	40	60	79	59

(*a*) Draw a scatter diagram.

(*b*) Draw the line of best fit.

(*c*) Jason scored 60 on paper 1 but was absent for paper 2. About how many marks would he have scored on paper 2, according to your diagram?   — Dosec c an lesson

3 The table gives the scores of eight golfers in a competition, together with their handicaps.

Handicap	3	4	7	10	12	12	13	18
Score	76	78	79	82	85	83	83	89

(*a*) Plot a scatter graph, and draw the line of best fit.

(*b*) Another golfer with a handicap of 15 entered the competition. What would he be expected to score, according to your diagram?   - ι lesson

4 The table lists the weight of, and the number of pages in, eleven books.

No. of pages	79	90	100	104	116	127	140	140	148	158	166
Weight (g)	168	190	190	212	242	273	275	301	330	340	359

(*a*) Draw a scatter graph and line of best fit.

(*b*) Another book has 135 pages. How heavy is it likely to be?

5 A numeracy test and a reading test was given to ten children in a junior school. The results are shown in the table.

Numeracy score	47	75	65	40	62	88	57	59	66	78
Reading score	24	38	30	23	32	48	30	36	34	41

(*a*) Draw a scatter graph and a line of best fit.

(*b*) Philippa has a reading score of 40. What would you estimate her numeracy to be, from your diagram?

Revision exercises: Chapters 5-10

Rewrite these questions, putting in brackets where necessary and working out the answer in each case.

1 $7 - 1 - 3$

2 $2 + 3 \times 6$

3 $12 - 4 + 1$

4 $15 \times 2 - 3 \times 4$

5 $40 \div 2 \times 3$

6 $10 + 2 \times 4 - 7$

Work out:

7 8^2

8 3^5

9 6^2

10 8^3

11 4^3

12 1^5

13 $2^4 \times 2^3$

14 $3^5 \div 3^2$

15 $4^5 \times 4^2$

16 $6^3 \div 6$

Write as ordinary numbers:

17 7.312×10^2

18 3.042×10^6

19 1.212×10^4

20 4.574×10^8

21 6.517×10^{-2}

22 8.32×10^{-5}

23 5.407×10^{-1}

24 8.309×10^3

Write the following numbers, shown on a calculator display, as ordinary numbers:

25 4.97   03

26 5.29 -02

27 7.21   04

28 1.123 -03

29 2.004   05

Write these numbers in standard form:

30 4630

31 324 012

32 501.04

33 76.3

34 0.0014

35 0.21

36 36 000 000

37 0.000 48

Work out:

38 3^{-2}

39 8^{-3}

40 3^0

41 10^{-3}

42 7^{-2}

43 1^{-4}

44 4^{-3}

45 2^1

Write these numbers correct to the number of decimal places stated:

46 4.942   (*a*) 1 d.p.   (*b*) 2 d.p.

47 52.417   (*a*) 1 d.p.   (*b*) 2 d.p.

48 0.482 97   (*a*) 1 d.p.   (*b*) 2 d.p.   (*c*) 3 d.p.

49 10.0309   (*a*) 2 d.p.   (*b*) 3 d.p.

50 91.9191   (*a*) 1 d.p.   (*b*) 2 d.p.   (*c*) 3 d.p.

51 1.0327   (*a*) 1 d.p.   (*b*) 2 d.p.   (*c*) 3 d.p.

52 14.4928   (*a*) 1 d.p.   (*b*) 2 d.p.   (*c*) 3 d.p.

Write these numbers rounded to the number of significant figures stated:

53 73.586   (*a*) 3 s.f.   (*b*) 4 s.f.

54 1.0507   (*a*) 2 s.f.   (*b*) 3 s.f.

55 0.004 27   (*a*) 2 s.f.   (*b*) 3 s.f.

56 174.25   (*a*) 3 s.f.   (*b*) 4 s.f.

57 538.42   (*a*) 1 s.f.   (*b*) 2 s.f.

58 488   (*a*) 1 s.f.   (*b*) 2 s.f.

59 1009.9   (*a*) 3 s.f.   (*b*) 4 s.f.

60 285.42   (*a*) 1 s.f.   (*b*) 2 s.f.

Work out:

61 3^2

62 $\sqrt{64}$

63 8^2

64 $\sqrt{9}$

65 $\sqrt{144}$

66 $\sqrt{0.25}$

67 13^2

68 $\sqrt{0.64}$

69 6^2

70 $\sqrt{4900}$

Find the length of the side *x* in each of these triangles:

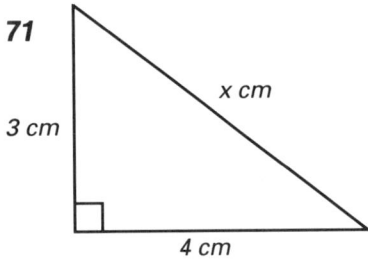

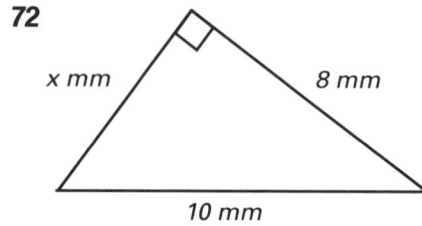

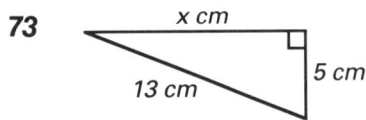

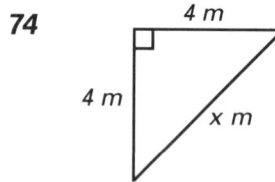

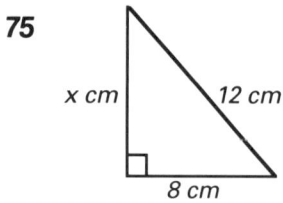

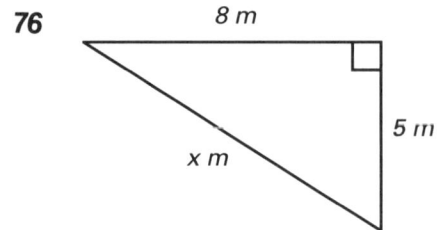

71

x cm

3 cm

4 cm

72

x mm

8 mm

10 mm

73

x cm

13 cm

5 cm

74

4 m

4 m

x m

75

x cm

12 cm

8 cm

76

8 m

5 m

x m

Insert either a $<$ or a $>$ between each pair of numbers to make a correct statement:

77 3 -5

78 5   0

79 -4 -2

80 3 -5

81 -6 -2

82 0   4

83 2 -2

84 -9 -1

Work out:

85 $3-4$	**86** $+2\times+3$	**87** $+8\div-2$	**88** $-3-(-2)$
89 $-4+7-9$	**90** $8-2$	**91** $-5\times+4$	**92** $-16\div-4$
93 $9-3$	**94** $-7\times+3$	**95** $-5+(-1)$	**96** $-4+(-2)$
97 -2×-4	**98** $-8-(-3)$	**99** $-20\div+5$	**100** $4-(-4)$
101 $+4\times-2$	**102** $7+(-7)$	**103** $+6\times-3$	**104** $-6-2-1$
105 $-3+7$	**106** $+8\times-2$	**107** $-4+3-9$	**108** $+8\div-2$
109 $-4-6$	**110** $(-4)^2$	**111** $-36\div-3$	**112** $-3+2-6$
113 $(-7)^2$	**114** -3×-10	**115** $-4-(-2)$	**116** $24\div-8$

117 A particular style of jewellery is made up of links of varying widths which can be joined in a variety of ways.

Link number	1	2	3	
Number of pieces	6			

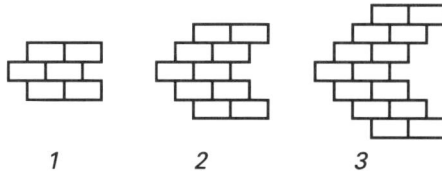

1   2   3

Complete the table to give the number of pieces in each link. Extend the table to find the number of pieces needed to make link numbers 4, 5 and 6.

Can you spot a pattern in the table? Find out how many pieces are needed for link numbers (*a*) 10 (*b*) 15 (*c*) 20.

118 A display is created by adding tennis balls as shown.

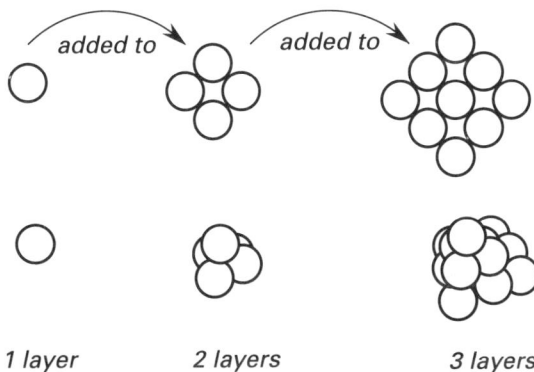

added to   *added to*

1 layer   2 layers   3 layers

Rows	1	2	3	4	5	6
No. of tennis balls	1	$1+4=5$	$5+9=$			

Complete the table for rows 4, 5 and 6.

How many tennis balls would be needed for a display of (*a*) 10 rows (*b*) 15 rows (*c*) 20 rows?

Tennis balls are sold in boxes of six. How many boxes would be needed for a display of (*a*) 10 rows (*b*) 15 rows (*c*) 20 rows?

119 Write in brackets the members of the following sets:

(*a*) {oceans of the world},

(*b*) {even numbers less than 20},

(*c*) {the colours of the rainbow},

(*d*){factors of 12}.

120 Describe in your own words these sets:

(*a*) {Zeus, Neptune, Thor, Woden},

(*b*) {Hotpoint, Zanussi, Hoover, Creda},

(*c*) {1, 4, 9, 16, 25},

(*d*){John Wayne, Jane Fonda, Michael J. Fox, Jane Seymour}.

121 How many members are there in each of these sets?

(*a*) {signs of the zodiac},

(*b*) {players in a netball team},

(*c*) {months in the year},

(*d*){multiples of 5 less than 50}.

122 Write down

(*a*) set *A*,

(*b*) set *B*,

(*c*) those numbers in both set *A* and set *B*,

(*d*) all the numbers in set $\mathcal{E}$,

(*e*) those numbers neither in set *A* nor in set *B*.

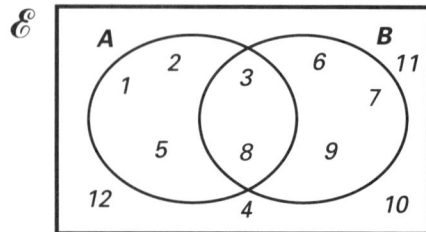

123 Write down

(*a*) set *X*,

(*b*) set *Y*,

(*c*) those letters in both set *X* and set *Y*,

(*d*) all the letters in set $\mathcal{E}$,

(*e*) those letters neither in set *X* nor in set *Y*.

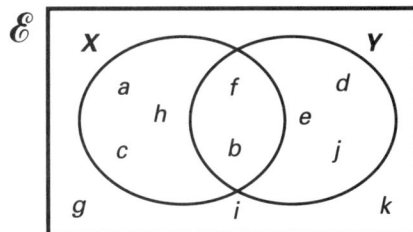

124 Write down

(*a*) all those who like just classical music,

(*b*) all those who like just pop music,

(*c*) all those who like either classical or pop music.

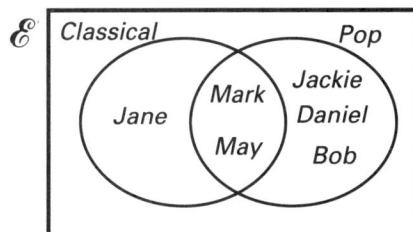

125 The numbers 1 to 16 are shown in the diagram. Write down

(*a*) all the factors of 12,

(*b*) all the factors of 16,

(*c*) all the numbers which are factors of both 12 and 16,

(*d*) all numbers which are not factors of either 12 or 16.

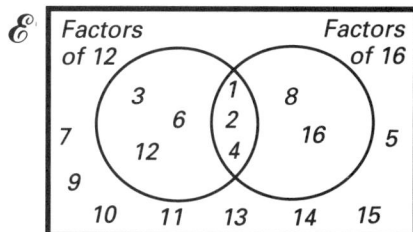

126 Draw a Venn diagram for each pair of sets:

(*a*) $P = \{1, 3, 5, 7, 9\}$ $\quad$ $Q = \{3, 4, 5, 6, 7\}$

(*b*) $X = \{a, b, c, d, e\}$ $\quad$ $Y = \{d, e, f, g, h\}$

(*c*) $D = \{\text{numbers less than 10}\}$ $\quad$ $E = \{\text{odd numbers less than 16}\}$.

127 Draw a Venn diagram for the sets in each question.

(*a*) Universal set $= \{\text{all numbers between 10 and 25}\}$,
$A = \{\text{factors of 24}\}$ $\quad$ $B = \{\text{multiples of 4}\}$.

(*b*) Universal set $= \{m, a, c, i, l, n, t, h, e, s\}$,
$C = \{a, c, e, i, l\}$ $\quad$ $D = \{l, t, h, s\}$.

(*c*) Universal set $= \{1, 2, 3, 5, 6, 7, 10, 11, 12, 14, 15\}$,
$P = \{3, 6, 12, 14, 15\}$ $\quad$ $Q = \{1, 3, 6, 10, 15\}$.

(*d*) Universal set $= \{\text{Haroon, Stephen, Debbie, Alex, Chris, Sadia, Anne}\}$,
$X = \{\text{Stephen, Haroon, Chris}\}$ $\quad$ $Y = \{\text{Debbie, Alex, Sadia}\}$.

128 The diagram shows the results of a survey taken in a particular street to find the type of car owned by each family.

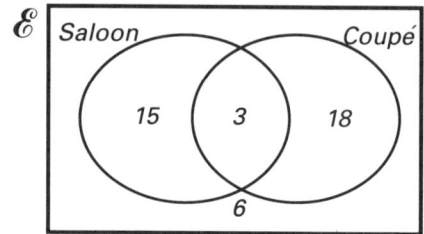

(a) How many families live in this street?

(b) How many have a saloon car?

(c) How many have a coupé car?

(d) How many families have a saloon and a coupé?

(e) Find how many families have no car.

129 A group of children were asked whether they had watched soccer or rugby on television last Saturday. The Venn diagram shows the results.

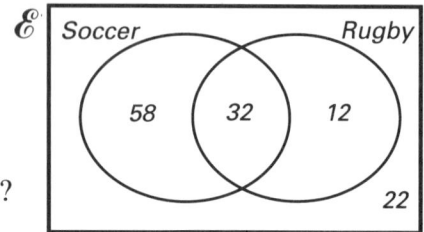

(a) How many had watched soccer?

(b) How many had watched both soccer and rugby?

(c) Find how many children were questioned.

(d) How many had watched neither soccer nor rugby?

(e) How many had not watched soccer?

130 Out of 50 children who were asked whether they had played hockey or netball during the last week, 12 had played netball, 28 had played hockey and 4 had played both.
Draw a Venn diagram to help you answer these questions.

(a) How many children had played neither hockey nor netball?

(b) How many had played just hockey?

(c) How many had played just netball?

(d) Find out how many children had not played hockey.

131 A third-year group is choosing options in humanities. A total of 74 chose history, 67 chose geography, and 32 chose a combined course of history and geography. The remaining 7 children in the year group did not choose a history or geography course.
     Draw a Venn diagram, and answer these questions:

(a) How many children are there in the year group?

(b) How many children have opted to follow a course which includes geography?

(c) How many children will study some history?

132 Write down the coordinates of the points A to H in the diagram.

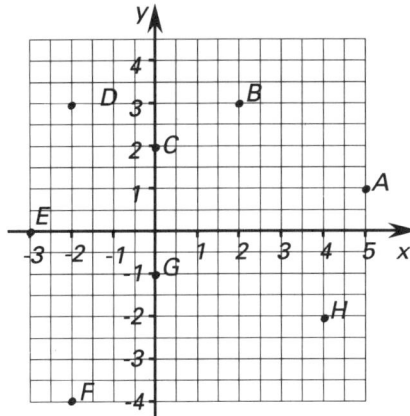

133 Mark the following points on a grid in which each axis ranges from −5 to +7:

A(4,6) B(1,7) C(−2,5) D(0,−4) E(−4,−3) F(−4,0)
G(6,−3)

134 Draw a grid, with the x axis ranging from −5 to 6, and the y axis ranging from −5 to 5.

(*a*) Join the points P(−2,4) and Q(6,−4) with a straight line.

(*b*) Join the points R(−2,1) and S(4,4) with a straight line.

(*c*) At which point do the lines PQ and RS meet?

(*d*) Write down the coordinates of three other points on line PQ. Can you work out a rule which connects the x and y coordinate of any point on the line PQ?

(*e*) Write down the coordinates of three more points on the line RS.

(*f*) Join the points T(3,−4) and U(6,2) with a straight line.

(*g*) At which point do the lines PQ and TU meet?

(*h*) By extending the lines in your diagram, state the coordinates of the point at which lines RS and TU meet.

135 Make up a table and draw the line represented by the equation $y = 2x + 5$. Take x values from 0 to 8.

136 Taking x values from −4 to +4, make a table of values and draw the line with equation $y = \dfrac{x}{2} + 3$.

Find (*a*) the y value when $x = 1.6$ (*b*) the x value when $y = 4.7$.

137 For the equation $y = 12 - x$, make up a table and draw the line, taking x values from -2 to 10.

   What is the value of y when $x = 4.5$?

138 Draw the line $y = 20 - 3x$, taking values of x from 0 to 8.
   (Remember that $20 - 3x = 20 + (-3x)$, when you are making up your table.)

139 Draw the line $y = 2(x + 1)$, for values of x from -4 to $+5$.

140 Make up a table for the curve $y = \dfrac{12}{x}$, for x values from 1 to 6.

   Plot the points, and join them with a smooth curve.
   From your graph, estimate:
   (a) the value of y when $x = 2.4$.

   (b) the value of x when $y = 3.6$.

141 Copy and complete the table, which gives the coordinates of points on the curve $y = x^2 - 3$.

x	0	1	2	3	4	5
x^2 -3		1 -3			16 -3	
$y = x^2 - 3$		-2			13	

   Plot the points, and draw a smooth curve through them.
   From your graph estimate the values of y when
   (a) $x = 3.5$ (b) $x = 0.5$.

142 Draw the graph of the equation $y = 3x^2$, making up a table of values, taking x from -4 to $+4$.
   From your graph, estimate:

   (a) the y value when $x = 2.5$,

   (b) the y value when $x = -4.2$,

   (c) the x values (both of them) when $y = 36$.

143 Work out the gradients of the lines AB, CD, and EF in the diagram.

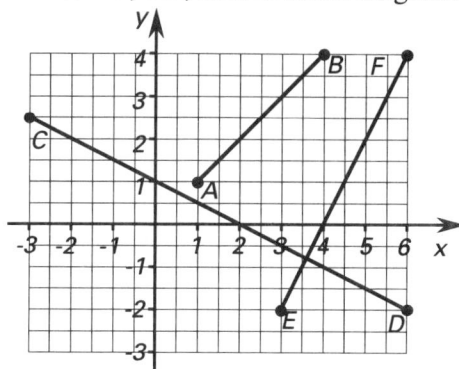

144 Plot the points P(1,4) and Q(3,8) on a grid.

(*a*) What is the gradient of the line PQ?

(*b*) Could you work out the gradient without drawing the line?

(*c*) What are the coordinates of the points where the line PQ cuts each axis?

(*d*) Another line has the same gradient as PQ, but passes through the point R(3,2). Draw this line on your diagram.

(*e*) Write down the coordinates of another point on this new line.

145 Three points L(−4,5), M(4,1) and N(−2,−4)) form a triangle. Work out the gradients of the sides LM, MN and NL of the triangle, by plotting the points, or otherwise.

Find also the gradient of the three lines OL, OM and ON, where O is the origin (0,0).

146 In each case, solve the pair of simultaneous equations by finding the point of intersection of the lines represented by each pair of equations:

(*a*) $y = x - 2$ and $y = 6 - x$

(*b*) $y = 2x$ and $y = \frac{x}{2} + 3$

(*c*) $y = 7 - x$ and $y = \frac{x}{3} + 3$

147 Using ruler and compasses, draw lines of length (*a*) 6 cm (*b*) 8.5 cm (*c*) 9 cm. Bisect each line.

148 Draw each of these three triangles accurately, using ruler and compasses:

(a)

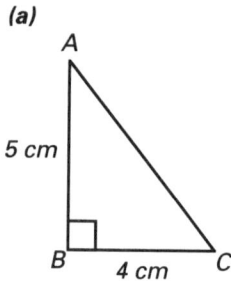

(b)

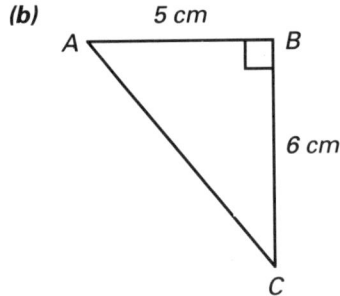

(c)

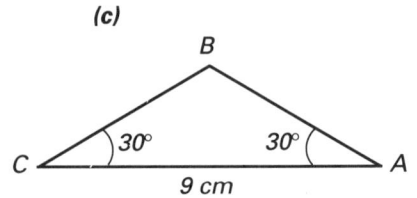

149 For each triangle in question 121: (*a*) bisect the line BC (*b*) bisect angle C.

150 Construct angles of 90°, 45°, 60° and 30° using only ruler and compasses.

151 Draw these shapes accurately, using ruler and compasses only.

(a)

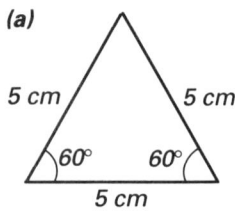

(b)

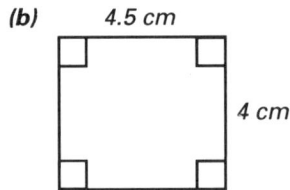

(c)

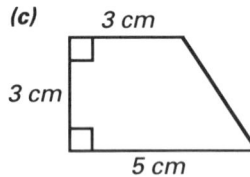

(d)

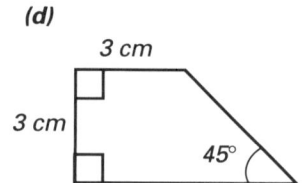

(e)

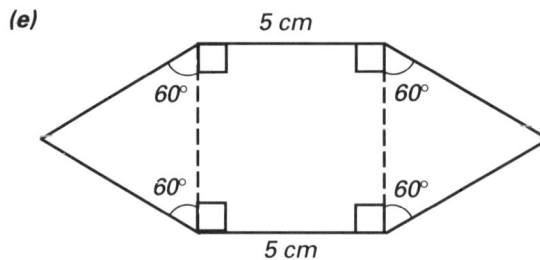

152 Describe and sketch the paths made by the following:

    (*a*) a stone thrown out from the top of a cliff,

    (*b*) a ball thrown into a net at basketball or netball,

    (*c*) an athlete's head while running in a hurdle race,

    (*d*) an aeroplane making a short flight between airports,

    (*e*) a ball caught in the top of a jet of water in a fountain.

153 ABCD is a horizontal rectangular enclosure. A ball is rolled from D, so that it is always the same distance from AD and DC. Draw the locus of the ball.

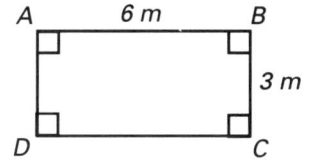

154 A bull is tethered at a point A by a chain of length 8 m, which is attached to the fence of a circular field of radius 10 m. Shade in the area of the field in which the bull can graze.

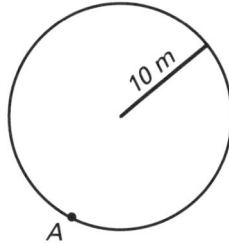

155 The diagram shows the position of three DIY superstores which operate a delivery service.

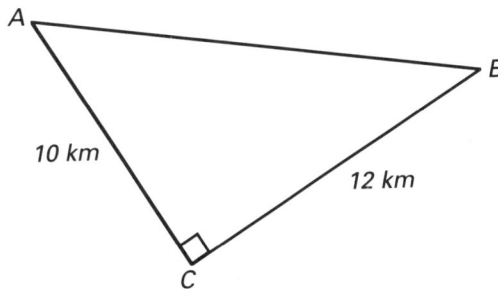

Store A will deliver to within 8 km of A.
Store B will deliver to within 10 km of B.
Store C will deliver to within 11 km of C.

By shading, show the region in which deliveries can be made from all three stores.

156 Copy the diagram, and show on it:

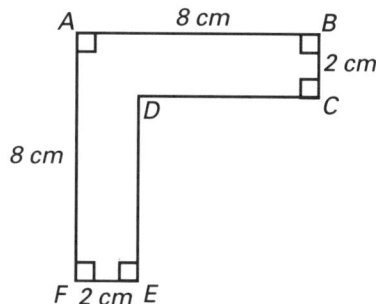

(*a*) the locus of all points equidistant from AB and CD,

(*b*) the locus of all points 0.5 cm from AF,

(*c*) the locus of all points equidistant from AB and AF.

157 Which of the shapes in the diagram are congruent with shape L (shaded)?

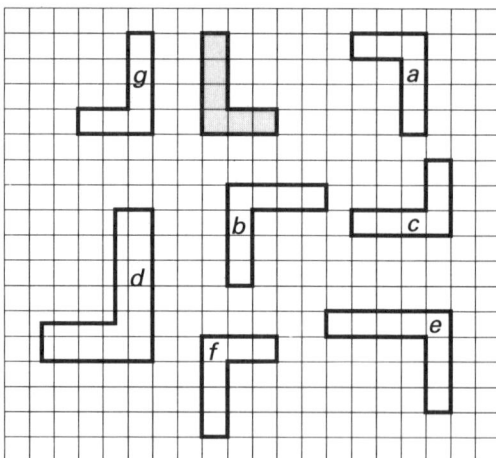

158 (*a*) Complete the diagram to make the second shape congruent with shape A.

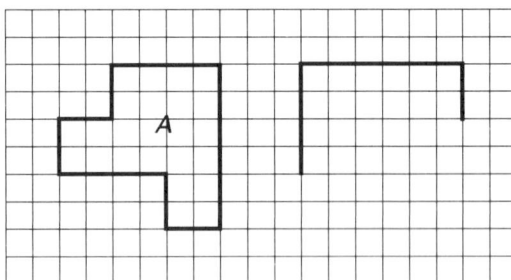

(*b*) Draw two more shapes congruent with A.

159 Draw triangle DEF, where D is (2,1), E is (2,5) and F is (−1,1).

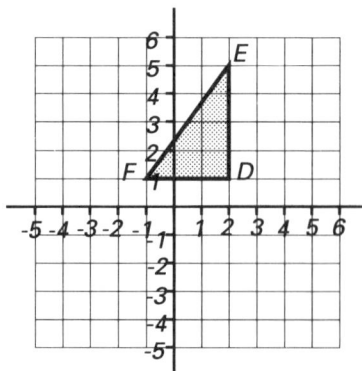

(*a*) Draw a triangle congruent with DEF, in which two of the corners are (6,2) and (3,2).

(*b*) Draw another congruent triangle, with two of the corners at (1,−3) and (5,−3).

(*c*) Draw another congruent triangle, this time with two corners at (−1,−5) and (−4,−1).

160 (*a*) Which of these shapes are simliar to shape M?

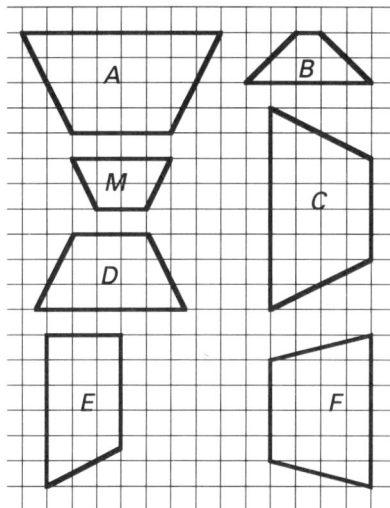

(*b*) Draw another shape similar to M, but with sides three times as long.

161 A photograph is enlarged so that the length of the enlarged photograph is $2\frac{1}{2}$ times the length of the original. The original measured 8 cm by 5 cm. What are the length and width of the enlarged photograph?

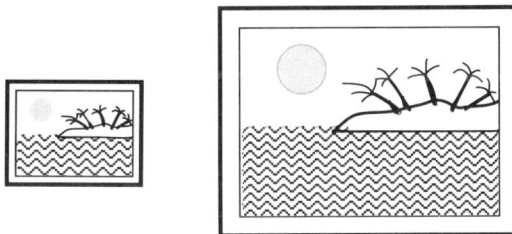

162 I wish to enlarge a photograph which is 17.5 cm long by 12.5 cm wide, so that it will fit on to a board 75 cm long. How wide will the board have to be, at least?

163 At a visit to an art gallery, I took a photograph of a painting. My photograph measures 12 cm by 12 cm. If the actual painting measures 96 cm by 96 cm, what is the scale factor of the enlargement?

164 A lady buys two skirts for her daughters. The skirts have waist measurements of 56 cm and 40 cm. If the skirts are similar, and the larger one is 84 cm long, how long is the smaller skirt?

165 Work out the angles marked by small letters in these diagrams:

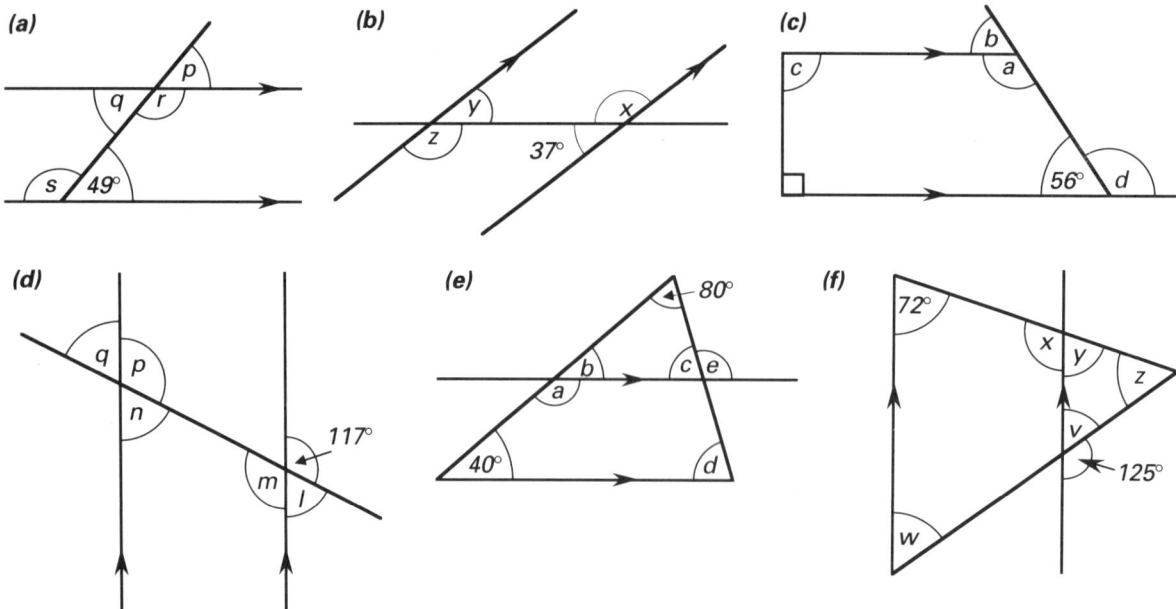

(a)

(b)

(c)

(d)

(e)

(f)

166 Work out the angles marked by small letters in these diagrams:

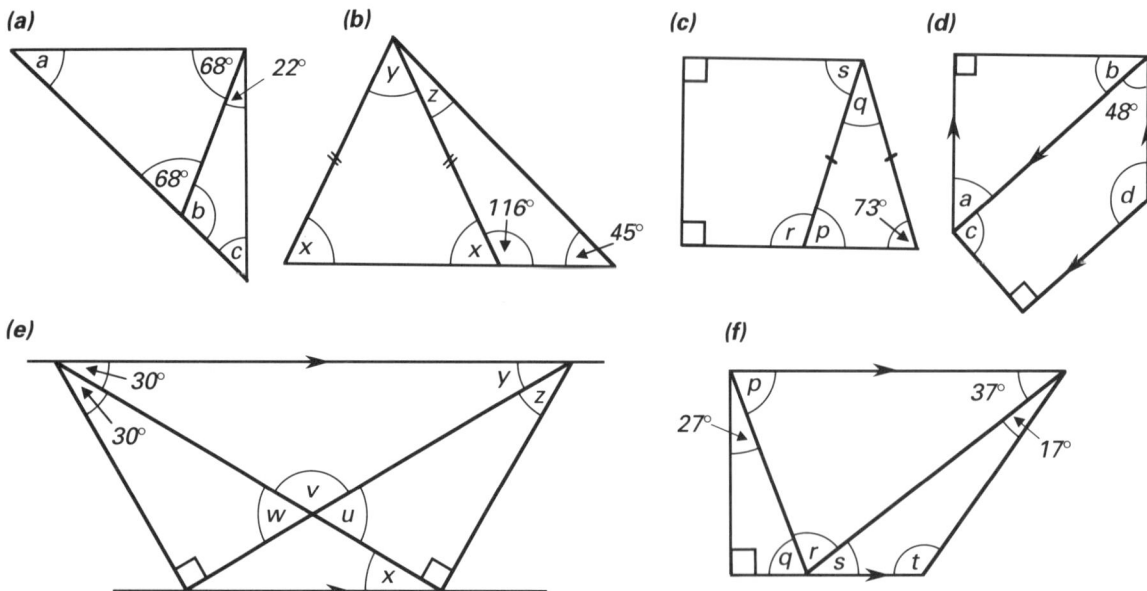

(a)

(b)

(c)

(d)

(e)

(f)

167 (*a*) Work out the sum of the angles in a pentagon.

   (*b*) In a sketch of a house end, the two angles at the bottom are each 90°, and the other three angles are all equal.

   Calculate the size of one of these angles.

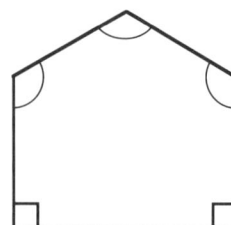

168 In this hexagon the two bottom angles are each 90°.

(*a*) If the other four angles are all equal, calculate the size of one of them.

(*b*) If instead, angles *p* and *q* are each 150°, calculate the size of *r* and *s*, given that they are equal to each other.

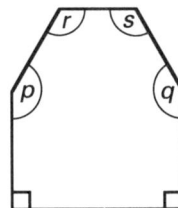

169 The diagram represents four sides of a regular twelve-sided polygon (dodecagon).

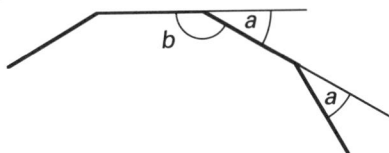

(*a*) By considering how many angles there are in one full turn, each equal to angle *a*, calculate angle *a*.

(*b*) Hence work out the size of the interior angle, *b*, of a regular dodecagon.

170 (*a*) Work out the sum of the angles in a hexagon.

(*b*) The diagram shows a square with two equilateral triangles.

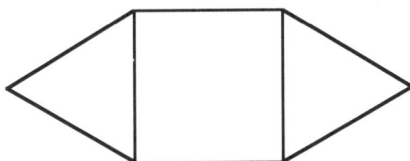

Check your answer to (*a*) from this shape.

171 In each of these diagrams work out the angles marked by letters:

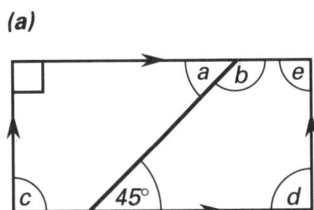

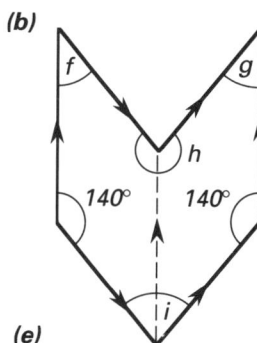

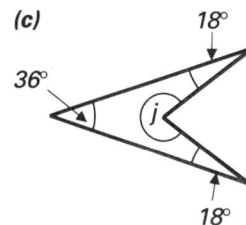

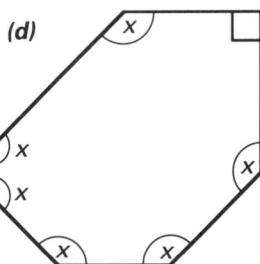

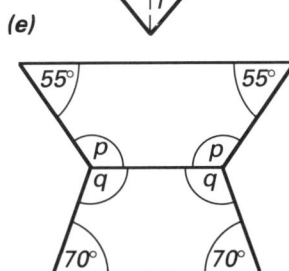

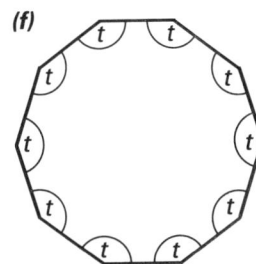

(*a*)

(*b*)

(*c*)

(*d*)

(*e*)

(*f*)

172 Use the fact that, in a circle the angle between a tangent and a radius is 90°, to work out the angles marked by letters in the following diagrams:

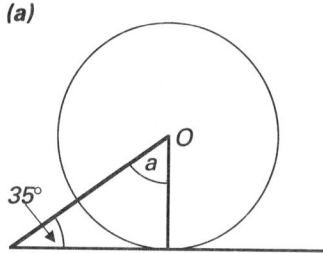

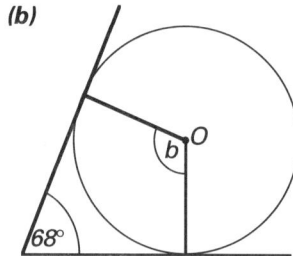

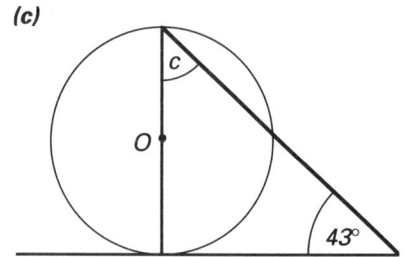

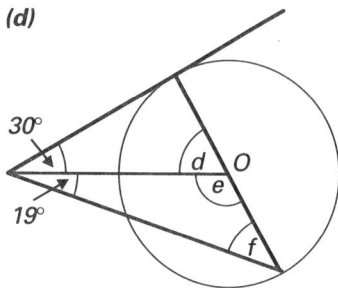

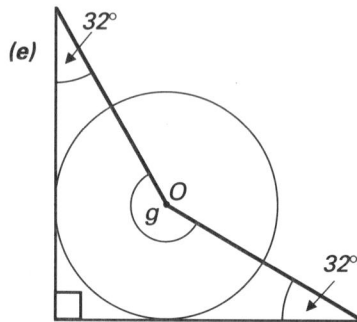

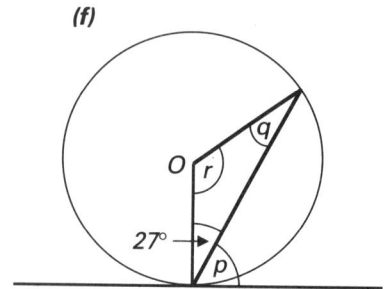

(a)

(b)

(c)

(d)

(e)

(f)

173 A youth club had a competition involving six teams. The total number of points scored by five of the teams is shown in the diagram (the red team's score is missing).

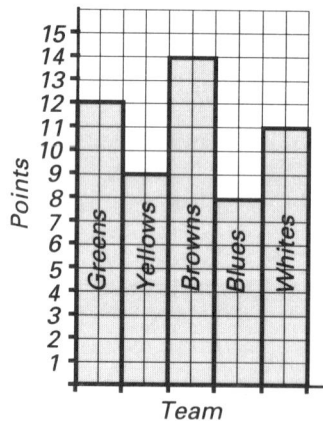

(*a*) How many points did the yellow team score?

(*b*) Which of these five teams scored most points?

(*c*) The teams are put into pairs. The scores of the yellow and brown teams are added. What is their total score?

(*d*) Another pair had the same total score. Which two teams?

(*e*) If the remaining team, together with the red team, had the highest score, what is the lowest that the red team could have scored?

174 The table gives the registration letter of cars taken in a survey for a mathematics project in July 1989, just before the 'G' registered cars were available.

Registration letter	X	Y	A	B	C	D	E	F
Number of cars	3	5	2	7	14	15	10	8

(*a*) Draw a bar chart of this information.

(*b*) How many cars were older than a 'B' registered car?

(*c*) How many cars were there altogether in the survey?

(*d*) How many cars were up to three years old?

(*e*) Mark says that 75% of the cars are 'C' registration or younger. Is he correct?

175 Ninety pupils were involved in an assessment in graphical communication, where the maximum mark was 12.

Test mark	0 1 2 3 4 5 6 7 8 9 10 11 12
No. of pupils	0 0 3 8 9 17 18 12 10 4 3 5 1

(*a*) Draw a frequency diagram of this information.

(*b*) How many pupils scored 4 marks or less?

(*c*) A grade C was awarded for 7 marks or more. How many pupils achieved at least a grade C?

(*d*) The top 10% were given a grade A. How many achieved a grade A? What mark did they have to reach in order to achieve a grade A?

(*e*) 'About half of the pupils scored 4, 5 or 6 marks.' Is this a reasonably correct statement?

176 The table shows the sales of a company during its first six years.

Year	1	2	3	4	5	6
Sales (£M)	3.4	4.0	5.2	5.7	6.9	8.4

(*a*) Plot the points, and draw a line graph through the points.

(*b*) During which year did the sales make less than the year before?

(*c*) Is it true that the sales in years 4 and 5 together were the same as in years 1, 2 and 3 together?

(*d*) Estimate the sales in year 7.

177 The noon temperature just outside the science laboratory was measured each day for a week.

Day	Mon.	Tue.	Wed.	Thu.	Fri.
Temp. (°C)	9.0	12.6	11.0	11.0	10.4

(a) Plot the points and draw a line graph.

(b) The temperature dropped by 1.6°C between which two days?

(c) What was the average noon temperature during the five days?

178 A lady decided to make a serious attempt to improve her golf, by having regular lessons. Every other month during this time she calculated her average score for a round of 18 holes.

Month	Feb.	Apr.	Jun.	Aug.	Oct.	Dec.
Average score	97	91	86	82	80	80

(a) Draw a line graph of her average scores.

(b) By how much had her average score fallen from February to December?

(c) Estimate her average score in March.

(d) When would her average score be 84?

179 The pie chart represents the weights of chemicals A, B, C and D in a mixture.

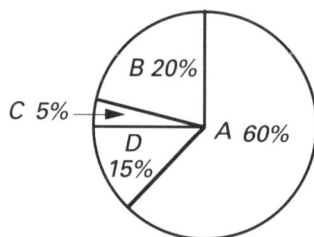

(a) What weight of chemical A is there in 100 g of mixture?

(b) I weigh out 300 g of mixture into a flask. How much of chemical C will there be in the flask?

(c) I have 10 g of chemical B. What weight of the mixture could I make if I used all 10 g of chemical B?

180 Three main courses were available at school dinner today: fish, ravioli, and tuna salad. The pie chart compares the numbers chosen of each meal.

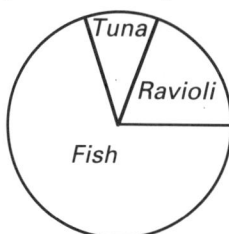

Four hundred main courses were chosen. How many were there of each of the three choices on offer?

181 Last year a market gardener worked out how much it had cost him for gas, electricity, coal and oil. The pie chart shows the proportions of each fuel used.

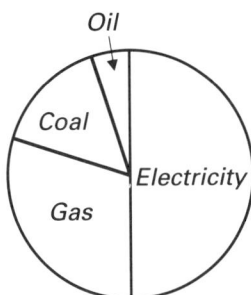

(*a*) What percentage of the cost was for coal and oil?

(*b*) His total bill came to £2000. How much did electricity cost him during the year?

(*c*) This year he expects to spend £900 on gas. Assuming that he uses fuel in the same proportions as last year, what are his total fuel costs likely to be?

182 The table shows the number of bulbs which I have, ready to plant.

Type of bulb	daffodil	crocus	tulip	hyacinth
Number	40	30	10	20

Draw a pie chart, labelling each sector.

183 The table shows the number of GCSE subjects entered by the 200 pupils in a fifth year.

Number of subjects	2 or less	3	4	5 or more
Number of pupils	50	30	40	80

Draw a pie chart of this information, taking care to label each sector.

184 Twenty-seven pupils completed an English assignment; the marks awarded were:

4	7	6	7	2	4	4	5	8
5	4	7	3	1	10	7	4	3
2	6	3	7	4	6	8	1	4

(a) Copy and complete the frequency table and tally column.

Mark	Tally	Frequency
0		
1		
2	I	
3		
4	I I	
5	I	
6	I	
7	I	
8		
9		
10		
Total		27

(b) Draw a bar chart of this information.

(c) 6, 7 or 8 marks were given a grade B. How many of these 27 pupils were given a grade B for their assignment?

185 In order to estimate the reading age of a written paragraph on a GCSE Design and Communication examination paper, the number of letters in each word of the paragraph is required. The list below gives the result of counting the number of letters in each word.

2	9	5	7	1	9	4	6	3	2
10	3	7	2	5	2	3	7	4	4
2	5	4	4	9	2	4	6	3	2
5	3	2	8	3	8	4	3	4	4
4	2	4	2	4	5	6	3	5	2
4	2	1	4	1	8	3	5	6	1
4	2	10	3	9	2	3	9	4	5

(*a*) Make up a frequency table, with a tally column.

(*b*) How many words are there in the paragraph?

(*c*) Draw a frequency diagram from your table.

(*d*) How many five-letter words are there?

(*e*) How many words contain seven letters or more?

186 At a 'pick-your-own' fruit farm, each basket of strawberries was weighed. During one afternoon the weights that were picked are shown in the frequency table.

Weight (kg)	from 1.0 up to 2.0	from 2.0 up to 2.2	from 2.2 up to 2.4	from 2.4 up to 2.6	from 2.6 up to 2.8	from 2.8 up to 3.0
Frequency	8	5	14	19	12	4

(*a*) Draw a frequency diagram to show the weight of strawberries picked.

(*b*) How many baskets weighed less than 2.2 kg?

(*c*) Estimate how many baskets weighed more than 2.7 kg.

(*d*) Assuming the eight lightest baskets each weighed 1.9 kg, what weight of strawberries was there altogether in these eight baskets?

(*e*) How much money would the farm collect on these eight baskets, if they charged 75p per kilogram?

187 Some students were employed as raspberry pickers. The weights, measured to the nearest $\frac{1}{10}$ of a kilogram, of 40 baskets of raspberries picked in one morning, is given below.

```
2.7 2.4 3.3 2.8 2.5 2.5 3.0 3.7 2.4 2.8
2.6 2.8 2.9 2.6 3.0 3.1 2.8 3.2 3.1 3.2
2.5 3.3 2.8 2.4 3.6 3.1 2.8 2.8 3.0 3.0
3.2 3.5 2.7 2.7 2.9 2.6 3.0 3.1 2.6 2.8
```

(*a*) Make up a frequency table, with a tally column. Begin with weights from 2.35 kg to 2.55 kg, then 2.55 kg to 2.75 kg, etc.

(*b*) Draw a frequency diagram.

(*c*) How many baskets weighed less than 2.75 kg?

(*d*) Estimate the number of baskets weighing over 3 kg.

(*e*) If the farm paid students 45p per kilogram picked, how much would a student be paid for picking 12.4 kg?

188 Find (*a*) the mode, (*b*) the median, of each set of figures:

(i) 5, 8, 8, 9, 10, 10, 10;

(ii) 16, 19, 17, 17, 11;

(iii) 3, 4, 6, 8, 9, 9;

(iv) 56, 47, 52, 56, 50, 48, 51, 48, 48, 50, 55.

189 (*a*) Find the mode, and the median, of these eight numbers: 4, 7, 3, 7, 10, 4, 7, 5.

(*b*) Another number is added to the list to make the mode and median equal. What is this number?

190 Find the lower quartile and the upper quartile of the list of numbers in question 188 part (iv).

191 In each part of question 188, work out the mean value.

192 The fifteen players on a rugby team are weighed. Their weights, in kilograms, are:

86, 72, 98, 95, 83, 88, 92, 94, 94, 92, 99, 78, 94, 90 and 95

Work out (*a*) the modal weight, (*b*) the median weight, (*c*) the lower and upper quartile weights, (*d*) the mean weight.

193 The two lightest players in question 192 above are replaced by two players who each weigh 97 kg. Work out:

(*a*) the new median weight,

(*b*) the new mean weight.

194 Draw a scatter graph for each part of this question. In each case draw the line of best fit.

(*a*)

x	3	3	4	5	6	7	8	9
y	1	2	2	4	5	5	6	8

(*b*)

x	22	22	24	25	26	26	27	28	30
y	5.2	5.4	5.6	5.6	5.9	5.8	5.9	6.2	6.1

(*c*)

x	13.4	13.6	13.7	14.0	14.3	14.5	14.6	14.7
y	3.3	3.4	3.6	3.7	3.7	3.7	3.8	3.8

Fractions: addition

A reminder

Example 1

Add together a half and a quarter.

You should remember that we can write $\frac{1}{2}=\frac{2}{4}$ (equivalent fractions).

1			
$\frac{1}{2}$		$\frac{1}{2}$	
$\frac{1}{4}$	$\frac{1}{4}$	$\frac{1}{4}$	$\frac{1}{4}$

So $\quad \frac{1}{2}+\frac{1}{4}=\frac{2}{4}+\frac{1}{4}=\frac{3}{4}$

(You do *not* add the 'tops' *and* the 'bottoms' – you must make the bottom numbers the same (e.g. quarters) by using equivalent fractions, then add the top numbers to see how many of these fractions you have. In this example you have 2 quarters and 1 quarter = 3 quarters = $\frac{3}{4}$.)

Example 2

Add $\frac{3}{8}$ and $\frac{1}{4}$.

Here we can change the $\frac{1}{4}$ into $\frac{2}{8}$.

So $\quad \frac{3}{8}+\frac{1}{4}=\frac{3}{8}+\frac{2}{8}=\frac{5}{8}$

1							
$\frac{1}{2}$				$\frac{1}{2}$			
$\frac{1}{4}$		$\frac{1}{4}$		$\frac{1}{4}$		$\frac{1}{4}$	
$\frac{1}{8}$	$\frac{1}{8}$	$\frac{1}{8}$	$\frac{1}{8}$	$\frac{1}{8}$	$\frac{1}{8}$	$\frac{1}{8}$	$\frac{1}{8}$

Example 3

In a village art competition, half of the prize money went to the best picture, a quarter went to the second best, and a sixth went to the third best. Celia's two pictures came first and third. What total fraction of the prize money did she win?

We need to add together $\frac{1}{2}$ and $\frac{1}{6}$. Remember that $\frac{1}{2}=\frac{3}{6}$, which means that we can add the fractions thus:

$$\frac{1}{2}+\frac{1}{6}=\frac{3}{6}+\frac{1}{6}=\frac{4}{6} \ (=\frac{2}{3})$$

So Celia won $\frac{2}{3}$ of the total prize money.

The process of finding appropriate **equivalent fractions** is essential in any work involved with combining fractions.

EXERCISE 11.1

As revision, try these simple questions by finding the missing numbers marked with an asterisk.

1 $\dfrac{3}{4} = \dfrac{*}{8}$ **2** $\dfrac{2}{5} = \dfrac{*}{10}$

3 $\dfrac{1}{4} = \dfrac{*}{12}$ **4** $\dfrac{4}{5} = \dfrac{*}{20}$

5 $\dfrac{2}{3} = \dfrac{*}{18}$ **6** $\dfrac{15}{20} = \dfrac{*}{4}$

7 $\dfrac{6}{10} = \dfrac{*}{5}$ **8** $\dfrac{10}{30} = \dfrac{*}{3}$

9 $\dfrac{12}{20} = \dfrac{*}{10} = \dfrac{*}{5}$ **10** $\dfrac{40}{60} = \dfrac{*}{6} = \dfrac{*}{3}$

Example 4

Add $\dfrac{1}{3}$ and $\dfrac{1}{4}$.

In this case, we cannot double or treble one of the bottom numbers (denominators) to make it equal to the other one, as we did in Examples 2 and 3 above.

We have to find a number into which *each* denominator will divide. In this example, both 3 and 4 divide into 12.

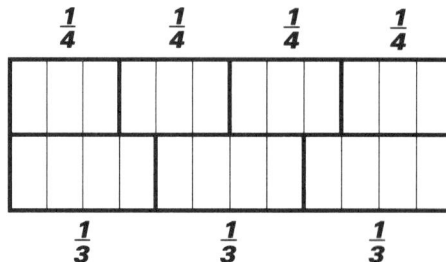

So $\dfrac{1}{3}$ can be changed into $\dfrac{4}{12}$, and $\dfrac{1}{4}$ into $\dfrac{3}{12}$. We then write:

$$\dfrac{1}{3} + \dfrac{1}{4} = \dfrac{4}{12} + \dfrac{3}{12} = \dfrac{7}{12}$$

Example 5

As part of my fitness programme I spend $\frac{3}{5}$ of my training time cycling and $\frac{1}{4}$ of my training time jogging. The rest of my training time I spend on brisk walks.

(*a*) For what fraction of the time do I either jog or cycle?

(*b*) What fraction of the time do I spend walking?

(*a*) I cycle or jog for $\frac{3}{5}+\frac{1}{4}$ of the time.

As $4 \times 5 = 20$, I can change each fraction into 20ths.

$$\frac{3}{5}=\frac{12}{20} \quad \text{and} \quad \frac{1}{4}=\frac{5}{20}$$

So $\frac{3}{5}+\frac{1}{4}=\frac{12}{20}+\frac{5}{20}=\frac{17}{20}$ of the time.

(*b*) The rest of the time must be $\frac{3}{20}$, as $\frac{3}{20}+\frac{17}{20}=\frac{20}{20}=1$, i.e. all of the time.

So I must spend $\frac{3}{20}$ of the time walking.

EXERCISE 11.2

Work out:

1 $\frac{1}{3}+\frac{1}{2}$ **2** $\frac{1}{5}+\frac{1}{3}$ **3** $\frac{2}{3}+\frac{1}{8}$

4 $\frac{1}{2}+\frac{2}{5}$ **5** $\frac{3}{8}+\frac{1}{3}$ **6** $\frac{1}{2}+\frac{1}{10}$

7 $\frac{5}{8}+\frac{5}{12}$ **8** $\frac{3}{7}+\frac{1}{2}$ **9** $\frac{1}{6}+\frac{3}{4}$

10 $\frac{1}{3}+\frac{1}{6}$

Example 6

Add together $\frac{3}{4}$ and $\frac{1}{2}$.

$\frac{1}{2}=\frac{2}{4}$, so $\frac{3}{4}+\frac{1}{2}=\frac{3}{4}+\frac{2}{4}=\frac{5}{4}$

The fraction $\frac{5}{4}$ however, can be looked upon as $\frac{4}{4}+\frac{1}{4}$, and $\frac{4}{4}$ is the same as one whole unit. So we can replace $\frac{5}{4}$ by $1\frac{1}{4}$.

($\frac{3}{4}$ of an hour plus $\frac{1}{2}$ an hour is one hour and a quarter.)

Example 7

Add together $3\frac{2}{3}$ and $5\frac{3}{4}$.

Here we add the whole number parts first, then the fraction parts. This gives us

$$8 + \frac{2}{3} + \frac{3}{4} = 8 + \frac{8}{12} + \frac{9}{12}$$

$$= 8 + \frac{17}{12}$$

$$= 8 + 1\frac{5}{12}$$

$$= 9\frac{5}{12}$$

EXERCISE 11.3

1 $\frac{1}{2} + \frac{4}{5}$

2 $\frac{3}{4} + \frac{1}{3}$

3 $\frac{2}{3} + \frac{1}{2}$

4 $\frac{3}{5} + \frac{3}{4}$

5 $\frac{5}{8} + \frac{5}{8}$

6 $2\frac{1}{3} + 4\frac{2}{5}$

7 $3\frac{1}{2} + 6\frac{1}{5}$

8 $2\frac{3}{4} + 4\frac{2}{3}$

9 $4\frac{7}{8} + \frac{4}{5}$

10 $\frac{1}{2} + 1\frac{5}{8}$

11 $\frac{1}{2} + \frac{1}{3} + \frac{1}{4}$

12 $3\frac{3}{8} + 4\frac{1}{4}$

13 $\frac{4}{7} + 2\frac{5}{7} + \frac{1}{2}$

14 $\frac{7}{20} + \frac{3}{4}$

15 $5\frac{1}{2} + \frac{17}{20}$

16 $\frac{2}{3} + \frac{3}{4} + \frac{4}{5}$

17 $4 + \frac{2}{3} + \frac{1}{6}$

18 $\frac{7}{10} + 1\frac{7}{10} + 2\frac{7}{10}$

19 $1\frac{5}{6} + 2\frac{1}{3} + 4\frac{1}{3}$

20 $12\frac{1}{2} + 1\frac{1}{12}$

Fractions: subtraction

Example 8

Work out $\frac{3}{4} - \frac{1}{3}$.

As with addition, we express each fraction with the same denominator (in this case 12), and then subtract the numerators. So

$$\frac{3}{4} - \frac{1}{3} = \frac{9}{12} - \frac{4}{12} = \frac{5}{12}$$

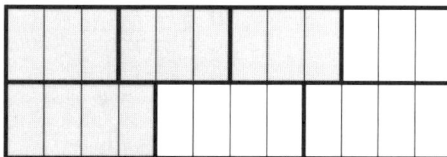

Example 9

Work out $5\frac{1}{2} - 2\frac{2}{5}$.

Dealing with the whole numbers first, $5 - 2 = 3$.

Then the fractions: $\frac{1}{2} - \frac{2}{5} = \frac{5}{10} - \frac{4}{10} = \frac{1}{10}$.

So $5\frac{1}{2} - 2\frac{2}{5} = 3\frac{1}{10}$.

Example 10

Work out $4\frac{1}{4} - 1\frac{2}{3}$.

The whole number part is $4 - 1 = 3$.
The fraction part is $\frac{1}{4} - \frac{2}{3} = \frac{3}{12} - \frac{8}{12} = -\frac{5}{12}$.

Together, we have 3 and $-\frac{5}{12}$.

As $1 - \frac{5}{12} = \frac{7}{12}$, $3 - \frac{5}{12}$ must be $2\frac{7}{12}$.

Hence $4\frac{1}{4} - 1\frac{2}{3} = 2\frac{7}{12}$.

EXERCISE 11.4

1 $\frac{5}{6}-\frac{1}{3}$

2 $\frac{3}{4}-\frac{1}{5}$

3 $\frac{1}{2}-\frac{1}{7}$

4 $\frac{2}{3}-\frac{1}{2}$

5 $\frac{4}{5}-\frac{3}{4}$

6 $\frac{9}{10}-\frac{3}{4}$

7 $\frac{6}{7}-\frac{2}{3}$

8 $\frac{7}{12}-\frac{1}{4}$

9 $\frac{8}{9}-\frac{3}{4}$

10 $\frac{3}{5}-\frac{3}{10}$

11 $5\frac{3}{5}-3\frac{1}{2}$

12 $6\frac{2}{5}-3\frac{1}{8}$

13 $4\frac{1}{2}-\frac{2}{5}$

14 $2\frac{1}{2}-1\frac{1}{3}$

15 $7\frac{7}{8}-4\frac{1}{4}$

16 $5\frac{1}{3}-1\frac{1}{2}$

17 $2\frac{1}{4}-1\frac{3}{5}$

18 $7\frac{2}{5}-4\frac{7}{10}$

19 $5-2\frac{3}{8}$

20 $6\frac{1}{6}-5\frac{5}{6}$

Investigation A

(a) From the set of increasing equivalent fractions $\frac{1}{3}, \frac{2}{6}, \frac{3}{9}, \frac{4}{12}$, etc., form a set of coordinates (3,1), (6,2), (9,3), (12,4), etc.

(b) Plot these points on a graph. What do you notice?

(c) Repeat, for different sets of equivalent fractions (e.g. $\frac{1}{2}, \frac{2}{4}, \frac{3}{6}$, and $\frac{3}{2}, \frac{6}{4}, \frac{9}{6}$).

(d) Draw the line through (1,1), (2,2), (3,3), etc.

(e) What conclusions can you draw about the lines on your diagram?

Fractions: multiplication

Example 11

Work out $\frac{2}{5}\times\frac{3}{4}$.

As you should remember, we multiply the numerators together, then the denominators, to give $\frac{2}{5}\times\frac{3}{4}=\frac{6}{20}$.

(We can simplify this fraction to its equivalent of $\frac{3}{10}$ if we wish.)

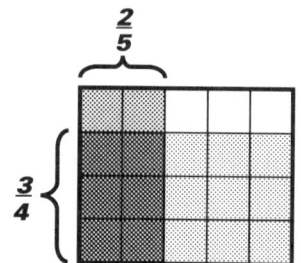

Example 12

Work out $3 \times \frac{4}{5}$.

Here we can look on 3 as the fraction $\frac{3}{1}$, which will give

$$3 \times \frac{4}{5} = \frac{3}{1} \times \frac{4}{5} = \frac{12}{5}$$

(We can write this as $2\frac{2}{5}$ if we wish.)

Example 13

Work out $2\frac{1}{4} \times \frac{2}{3}$.

In the first fraction there is a whole number part as well as a numerator and denominator. In cases like this we *must* make the fraction 'top-heavy' first, before we start multiplying.

So the $2\frac{1}{4}$, or $2 + \frac{1}{4}$, becomes $\frac{8}{4} + \frac{1}{4} = \frac{9}{4}$, and the calculation becomes

$$2\frac{1}{4} \times \frac{2}{3} = \frac{9}{4} \times \frac{2}{3} = \frac{18}{12} = 1\frac{6}{12} = 1\frac{1}{2}$$

Example 14

Work out $3\frac{1}{3} \times 1\frac{2}{5}$.

Here we make both fractions 'top-heavy':

$$3\frac{1}{3} \times 1\frac{2}{5} = \frac{10}{3} \times \frac{7}{5} = \frac{70}{15} = 4\frac{10}{15} = 4\frac{2}{3}$$

EXERCISE **11.5**

1 $\frac{3}{5} \times \frac{1}{2}$

2 $\frac{2}{3} \times \frac{3}{4}$

3 $\frac{5}{7} \times \frac{1}{4}$

4 $\frac{2}{3} \times \frac{2}{3}$

5 $\frac{1}{2} \times \frac{2}{3}$

6 $\frac{3}{5} \times \frac{9}{10}$

7 $\frac{3}{4} \times \frac{5}{6}$

8 $\frac{8}{11} \times \frac{1}{4}$

9 $\frac{3}{8} \times \frac{5}{6}$

10 $4 \times \frac{7}{8}$

11 $\frac{4}{5} \times 6$

12 $1\frac{3}{4} \times \frac{2}{3}$

13 $5\frac{1}{2} \times \frac{1}{4}$

14 $\frac{3}{7} \times 1\frac{1}{5}$

15 $3\frac{1}{2} \times 3\frac{1}{2}$

16 $\frac{1}{2} \times \frac{2}{3} \times \frac{3}{4}$

17 $2\frac{2}{5} \times 1\frac{1}{4}$

18 $\frac{3}{8} \times 2\frac{2}{3}$

19 $3\frac{1}{3} \times 6\frac{2}{3}$

20 $1\frac{3}{7} \times 1\frac{2}{5}$

Investigation B

(a) Are the answers to $\frac{3}{8} \times \frac{5}{7}$ and $\frac{3}{7} \times \frac{5}{8}$ the same?

(b) Are the answers to $\frac{a}{b} \times \frac{c}{d}$ and $\frac{a}{d} \times \frac{c}{b}$ always the same?

(c) Use this process to simplify the calculation of $\frac{5}{7} \times \frac{7}{10}$.

(d) Choose other examples in which a and d have factors, and b and c have factors, so that the calculation is made simpler.
(You may like to try these, as practice:

$$\frac{4}{5} \times \frac{5}{8}, \frac{7}{3} \times \frac{3}{7}, 4\frac{1}{2} \times \frac{2}{3}.)$$

Fractions: division

Example 15

Work out $\frac{2}{3} \div \frac{1}{5}$.

In division we proceed as in adding, by finding a denominator into which 3 and 5 will divide; 15 in this case. So we have

$$\frac{2}{3} \div \frac{1}{5} = \frac{10}{15} \div \frac{3}{15}$$

This is now 10 of these '15ths' divided by 3 of these '15ths', which is just $\frac{10}{3}$ or $3\frac{1}{3}$.

So once you have made the denominators equal (15 in this example) you then divide the *numerators*, $10 \div 3$, to give the answer.

Example 16

Work out (a) $\frac{1}{3} \div \frac{3}{4}$ (b) $7 \div \frac{2}{5}$.

(a) The 'common' denominator is 12.
So $\frac{1}{3} \div \frac{3}{4} = \frac{4}{12} \div \frac{9}{12} = \frac{4}{9}$.

(b) $7 \div \frac{2}{5} = \frac{7}{1} \div \frac{2}{5}$

$$= \frac{35}{5} \div \frac{2}{5} = \frac{35}{2} = 17\frac{1}{2}.$$

Example 17

Work out (a) $\frac{5}{8} \div 3$ (b) $4\frac{1}{2} \div 3\frac{1}{3}$.

(a) $\frac{5}{8} \div 3 = \frac{5}{8} \div \frac{3}{1}$

$= \frac{5}{8} \div \frac{24}{8} = \frac{5}{24}.$

(b) $4\frac{1}{2} \div 3\frac{1}{3} = \frac{9}{2} \div \frac{10}{3}$

$= \frac{27}{6} \div \frac{20}{6} = \frac{27}{20} = 1\frac{7}{20}.$

EXERCISE 11.6

1 Jo takes $\frac{1}{2}$ of a cake and Les takes $\frac{1}{4}$. What fraction of the cake has been taken? What fraction remains?

2 I need $\frac{1}{4}$ of a kilogram of flour for biscuits, and $\frac{2}{3}$ of a kilogram of flour for buns. What fraction of a kilogram bag of flour do I need altogether?

3 When Peter and I were practising our golf drives, Peter estimated that I hit my first drive $\frac{1}{3}$ of the way towards the hole, and my second drive half way towards the hole. By what fraction of the total distance to the hole was my second drive further than my first?

4 A screw, $\frac{5}{8}$ inch long, passes through one piece of wood $\frac{3}{8}$ inch thick, and is screwed tightly into a second piece of wood. How far into the second piece of wood does the screw penetrate?

5 A layer of hardboard, $\frac{1}{4}$ inch thick, is attached to a piece of chipboard $\frac{5}{8}$ inch thick. How thick are both pieces of board together?

6 Rashid and Ben have won their school caps at cricket. Rashid's cap size is $6\frac{3}{4}$ inches, while Ben's is $7\frac{1}{8}$ inches. How much larger is Ben's cap size than Rashid's?

7 I have a box of nails each of length $2\frac{1}{5}$ cm. Four of these nails, placed end to end, just reach from one side to the other of a block of wood. How wide is the block of wood?

8 Seven nails, each of length $1\frac{3}{5}$cm, when placed end to end just reach along the length of a block of wood. How long is the block of wood?

9 One of the $2\frac{1}{5}$ cm nails and one of the $1\frac{3}{5}$ cm nails are placed end to end. What distance do they cover?

10 A nail of length $2\frac{1}{2}$ cm is hammered through a block of wood $1\frac{4}{5}$ cm thick. How far through the block does the nail protrude?

11 At an all-ticket football match, $\frac{1}{3}$ of the tickets were sold to supporters of the home team, and $\frac{1}{5}$ were sold to supporters of the away team.

  (*a*) What fraction of the tickets were sold to both home and away supporters?

  (*b*) If there were 30 000 tickets sold, how many went to the home supporters?

12 Four neighbours decide to have gravel put on to the lane which runs between their houses. Mr Whittaker agrees to pay $\frac{2}{5}$ of the cost, and Mrs Henderson agrees to pay $\frac{1}{4}$ of the cost.

  (*a*) What fraction of the total cost do Mr Whittaker and Mrs Henderson pay?

  (*b*) What fraction do the other two neighbours pay?

  (*c*) If the total cost is £200, how much is due to be paid by (i) Mr Whittaker, (ii) Mrs Henderson, (iii) the other two neighbours?

13 Sandra and Teresa have a length of ribbon and each girl is going to cut a piece from it. Sandra cuts off $\frac{1}{4}$ of the ribbon and passes the remaining piece to Teresa. Teresa then cuts $\frac{1}{3}$ off the remaining piece.

  (*a*) Which girl has the longer piece?

  (*b*) If the ribbon is 8 m long initially, how long is each girl's piece, and how long is the piece that remains?

14 How many pieces, each $\frac{3}{5}$ m long, can I cut from a rope $2\frac{2}{5}$ m long?

15 There were three finalists in a school quiz. Of the 40 questions which were asked, Chi answered 20, Fiona answered 15 and Margaret answered the other 5.

(*a*) What fraction of the questions did Fiona answer?

(*b* Margaret answered all her questions correctly, but the other two *each* made four mistakes. What fraction of the 40 answers were given correctly?

16 Can Jane, Jean and Joan share a cake so that Jane has $\frac{1}{2}$, Jean has $\frac{1}{3}$ and Joan has $\frac{1}{4}$?

17 My aunt is complaining that her water rates are $2\frac{1}{2}$ times the amount they were five years ago, yet they are due to rise next month by $\frac{1}{5}$. What multiple will the new rate be of the rate five years ago?

18 The school examination officer has a box of calculators. Of these, $\frac{1}{5}$ are solar powered, and of the others, $\frac{1}{4}$ need new batteries.

(*a*) If there are 40 calculators in the box, how many of the battery-powered ones work?

(*b*) What fraction of the total need new batteries?

19 Amanda takes $\frac{2}{3}$ of the buttons from a button box. She then gives Barbara $\frac{1}{4}$ of these buttons. Caroline says that Barbara now has $\frac{1}{6}$ of the original total.

(*a*) Is Caroline correct?

(*b*) How many buttons will Barbara have if there were 60 buttons in the box initially?

20 How many glasses of lemonade, each holding $\frac{1}{8}$ litre, can you pour from a bottle containing $1\frac{1}{2}$ litres?

21 My bedroom floor measures $3\frac{1}{3}$ yards by $2\frac{1}{2}$ yards. What area of carpet, measured in square yards, do I need to cover the floor?

Decimals: applications

Here are three examples of practical applications of decimals.

Example 18

Which fraction is greater, $\frac{4}{11}$ or $\frac{7}{18}$?

A simple way of comparing two fractions is to change each into its decimal equivalent.

From your calculator, $\frac{4}{11} = 0.363\,636\ldots$ and $\frac{7}{18} = 0.388\,88\ldots$

Hence $\frac{7}{18}$ is greater than $\frac{4}{11}$.

Example 19

My bedroom floor measures 2 m 85 cm by 3 m 17 cm. How much will it cost me to cover the floor with a carpet costing £12.99 per square metre?

The length and width of my bedroom can be written in metres as decimals, then multiplied to give the area in square metres.

Area $= 2.85\,\text{m} \times 3.17\,\text{m} = 9.0345\,\text{m}^2$.

The cost of the carpet is therefore £12.99 $\times$ 9.0345 = £117.35816. To the nearest penny, the carpet costs £117.36.

Example 20

Yesterday my uncle drove me to see my sister. It took us 2 hours and 35 minutes to travel the 206 km from my house to my sister's house. What was our average speed?

$$\text{Average speed} = \frac{\text{total distance covered}}{\text{total time taken}} = \frac{206\,\text{km}}{2\,\text{h}\,35\,\text{min}}$$

The problem is the 2 h 35 min; one solution is to write it as a *decimal*.

Now 35 minutes is $\frac{35}{60}$ of an hour. Your calculator will give you this as 0.583333.... The time taken was therefore 2.583...hours. Hence our average speed was

$$\frac{206}{2.58333} = 79.74\ldots \text{ or } 80\,\text{km/h to the nearest km/h.}$$

Investigation C

A new window in my room needs three rectangular pieces of glass. The measurements are 410 mm by 250 mm, 410 mm by 570 mm and 450 mm by 830 mm. The man at my local hardware shop sells the glass I require for £1.75 per square foot.

Use the information below to work out how much I will have to pay for the three panes of glass.

$1000\,\text{mm} = 1\,\text{m}$ $\qquad 1\,\text{m}^2 = 10.76$ square feet

EXERCISE **11.7**

1 Which is the larger fraction, $\frac{2}{3}$ or $\frac{13}{20}$?

2 My brother is having a new carpet in his bedroom, which measures 2.52 m by 3.69 m.

   (*a*) What area of carpet is required to cover the whole floor?

   (*b*) How much will the carpet cost, at £10.50 per square metre?

3 Duncan swims four lengths of the swimming pool in these times: 53.5 s, 54.6 s, 56.2 s and 52.7 s.

   (*a*) What is the total time for the four lengths?

   (*b*) What is the difference between his fastest and slowest times?

4 Fran and Jeff are arguing about who had the better average score when they were at their rifle club. Fran hit the target 16 times out of 26 shots, Jeff hit the target 9 times out of 15 shots. Who had the better average score?

5 In our region, not all rugby teams play an equal number of games in a season. The championship is given to the team with the highest average number of points per game. At the end of last season the three top teams were: Lions with 106 points from 11 games, Panthers with 157 points from 16 games and Jaguars with 188 points from 19 games.

   (*a*) Which team won the championship?

   (*b*) Which team came second?

6 Which is better value for money, a 75 cl bottle of wine for £2.19 or a 1 litre bottle for £2.98?

7 For a stall at the School Fair I buy a box of 24 cans of orange juice for £3.99 and another box of 15 cans for £2.89. What is the average cost of one of these cans? Give your answer correct to the nearest penny.

8 On paper 1 of a test, I scored 84 out of a possible 137, and on paper 2, I scored 33 out of 50. On which paper did I do better?

9 I ran a 400 metre race in 53 seconds. What was my average speed in (*a*) metres per second (*b*) metres per hour (*c*) kilometres per hour?

10 A train leaves a station at 6.47 a.m. and is due to arrive at 9.53 a.m. at its destination, having travelled 273 miles.

   (*a*) How long is the journey, in hours and minutes?

   (*b*) If the train arrives on time, calculate its average speed, in miles per hour.

12. Measuring: shapes and solids

Area and perimeter

In Book 3X, Chapter 13 some common shapes were dealt with and you should now be familiar with these:

area of a rectangle = length × breadth

A rectangle can be divided into two equal triangles. It can be deduced that:

area of a triangle = $\frac{1}{2}$(base × height)

The **perimeter** of a shape is the distance all around the edge.

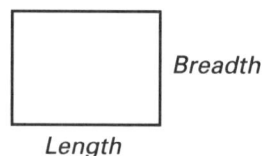

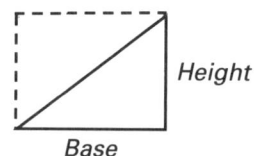

EXERCISE 12.1

For each of the following find (*a*) the perimeter (*b*) the area.

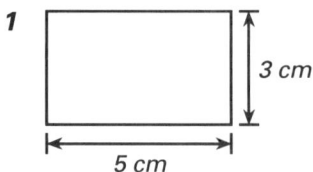

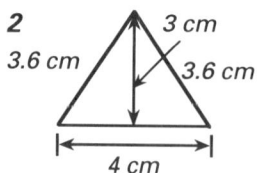

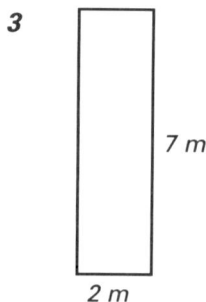

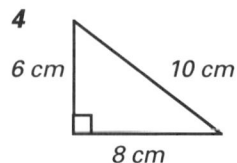

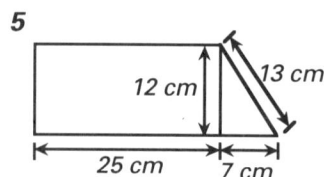

1 3 cm, 5 cm

2 3 cm, 3.6 cm, 3.6 cm, 4 cm

3 7 m, 2 m

4 6 cm, 10 cm, 8 cm

5 12 cm, 13 cm, 25 cm, 7 cm

For each of the following find the total area.

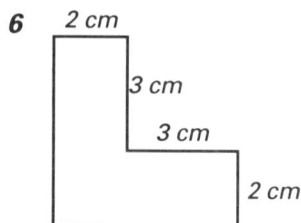

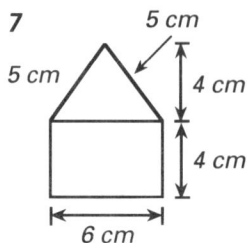

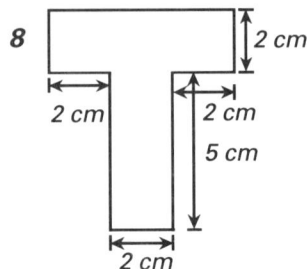

6 2 cm, 3 cm, 3 cm, 2 cm

7 5 cm, 5 cm, 4 cm, 4 cm, 6 cm

8 2 cm, 2 cm, 2 cm, 5 cm, 2 cm

9 Find the area of the shaded border.

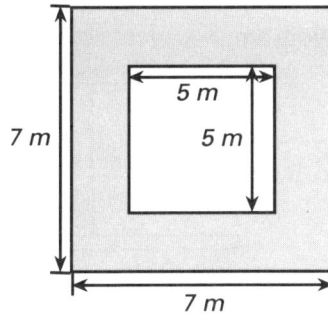

10 Find the area of the room shown.

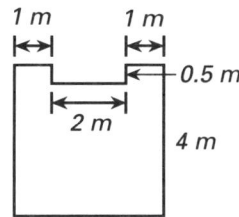

Area of a parallelogram

How can we find the area of this parallelogram? Trace the parallelogram on to a piece of paper and cut it out. Next cut down the line as shown.

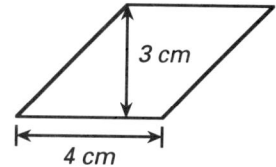

Move this triangle over to the other side of the parallelogram.

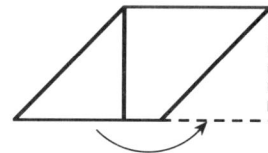

We have made a rectangle. Since we have not changed or removed any part of the area of the shape this shows us that the area of the parallelogram is the same as the area of the rectangle.

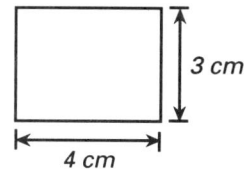

Area of rectangle = area of parallelogram
= base × height = 4 × 3 = 12 cm²

Example 1

Find the area of the parallelogram.

Area = 8 × 5 = 40 cm²

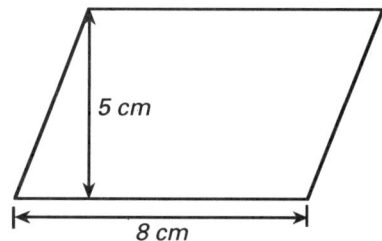

EXERCISE **12.2**

Find the area of each parallelogram.

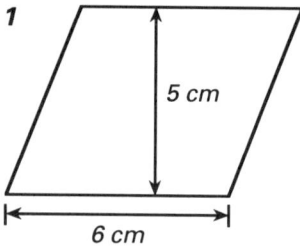

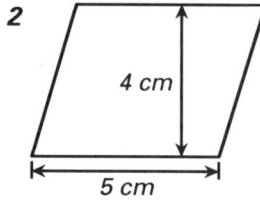

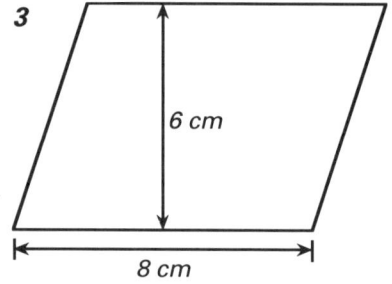

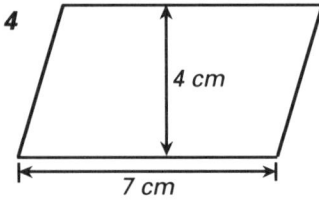

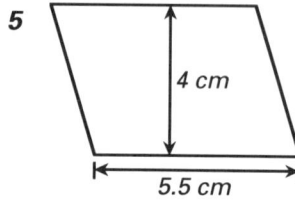

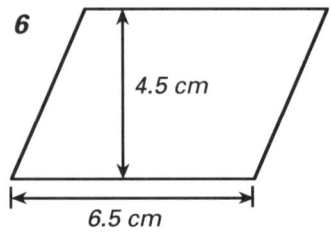

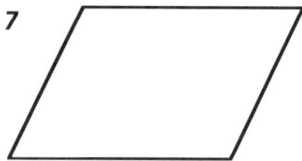

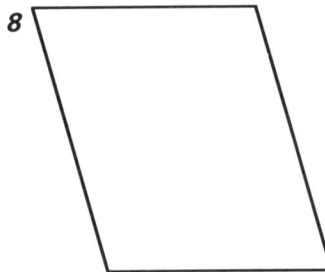

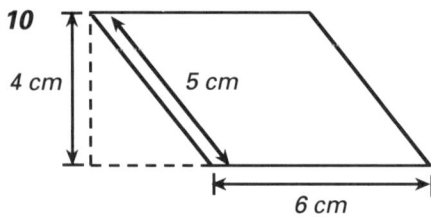

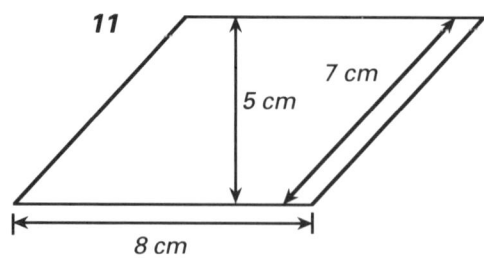

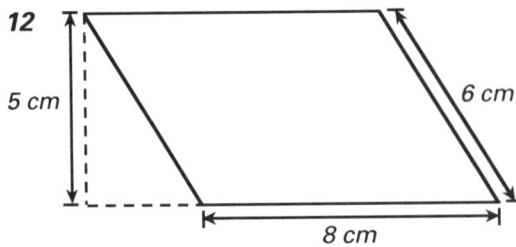

1

5 cm

6 cm

2

4 cm

5 cm

3

6 cm

8 cm

4

4 cm

7 cm

5

4 cm

5.5 cm

6

4.5 cm

6.5 cm

7

8

9

10

4 cm

5 cm

6 cm

11

5 cm

7 cm

8 cm

12

5 cm

6 cm

8 cm

EXERCISE **12.3**

1   Find the area.

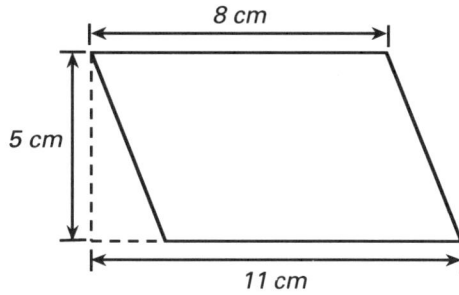

2   Find (*a*) the area (*b*) the perimeter.

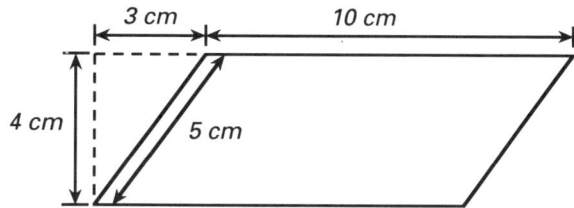

3   Find (*a*) the area (*b*) the perimeter.

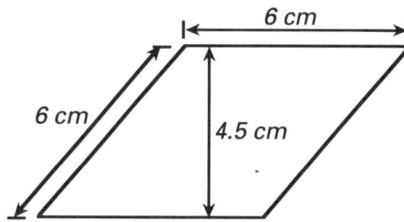

Draw these accurately and take measurements to help you find the area.

4

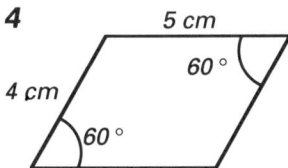

5

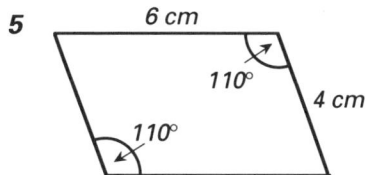

6

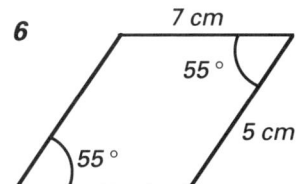

7   Find the total area.

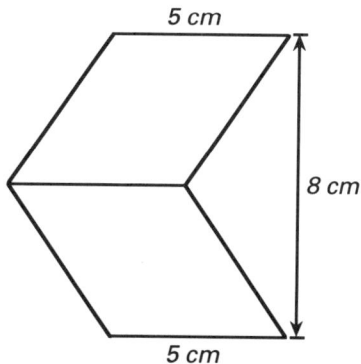

8 Find the total shaded area of this shape.

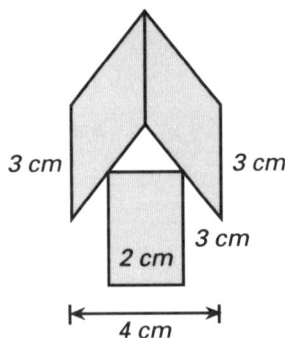

3 cm   3 cm

3 cm

2 cm

4 cm

Circumference and area of a circle

In Book 3X, Chapter 13 we found that the circumference of a circle was slightly more than three times as long as the diameter of the circle, and we used the fact that the circumference was given by:

circumference $= \pi \times$ diameter

where π was approximately 3.142, or the value given by a calculator with a π button or function. (In the following work π is taken from the calculator.)

Example 2

Find the circumference of the circle.

2 cm

If the radius is 2 cm, then the diameter is 2×2 cm $= 4$ cm.
Circumference $= \pi \times$ diameter $= \pi \times 4 = 12.566$ cm.

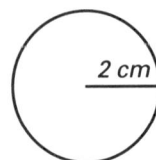

Investigation A

The area of a circle is the amount of space inside the circle.
   Draw a circle radius 2 cm on a sheet of graph paper and count the number of squares inside it. Fractions of squares need to be added together to make up whole squares. This is one way we can find the area enclosed within the circle. Do you think this is a good method for measuring the area of a circle?

Investigation B

Draw a circle of radius 2 cm again, cut it out, and cut up the circle into sectors as shown.

Place these sectors together:

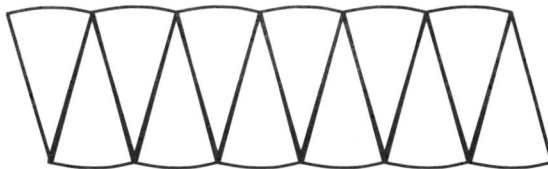

Take the end sector, cut it in half, and remove one half to the other side to give us a more familiar shape:

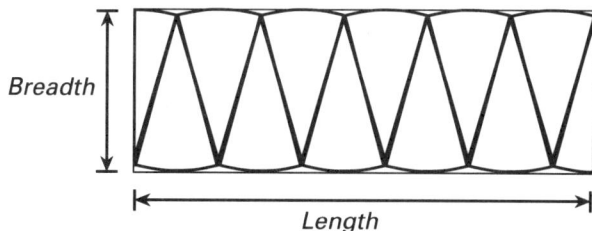

The shape we have made from the circle looks like a rectangle. We could find the area of the rectangle easily if we knew its length and breadth. Measure them; the breadth should be about 2 cm, as this is the radius of the circle, and the length should be about 6.3 cm, half the circumference of the circle, or $\pi \times r$.

The area of the rectangle is then $2 \times 6.3 = 12.6 \text{ cm}^2$, or $\pi \times r \times r = \pi r^2$. To find the area of any circle we use:

$$\text{area} = \pi r^2 = \pi \times \text{radius} \times \text{radius}$$

Example 3

Find the area of the circle.

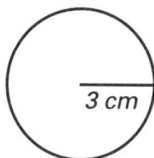

$$\begin{aligned} \text{Area} = \pi r^2 &= \pi \times \text{radius} \times \text{radius} \\ &= \pi \times 3 \times 3 = 28.274 \text{ cm}^2 = 28.3 \text{ cm}^2 \text{(to 3 s.f.)} \end{aligned}$$

Example 4

Find the area of the circle.

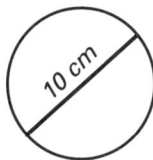

The diameter is 10 cm, so the radius is $10 \div 2 = 5$ cm.

$$\begin{aligned} \text{Area} = \pi r^2 &= \pi \times \text{radius} \times \text{radius} \\ &= \pi \times 5 \times 5 = 78.54 \text{ cm}^2 = 78.5 \text{ cm}^2 \text{(to 3 s.f.)} \end{aligned}$$

EXERCISE **12.4**

Find the area of each of the following circles, writing your answers to 3 significant figures.

1 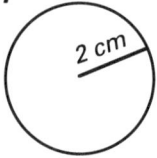   **2** 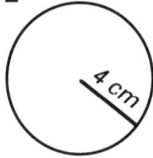   **3** 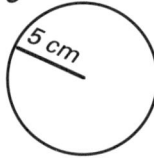   **4** 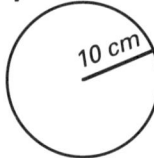   **5**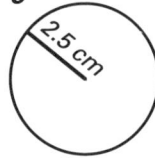

Write down the length of the radius of each of these circles, and find the area.

6 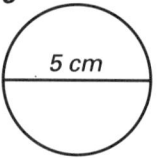   **7** 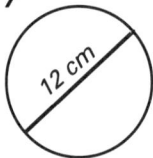   **8** 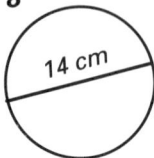   **9** 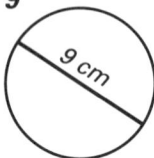   **10**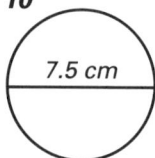

Find the area of these circles, writing your answers to 3 significant figures.

11    **12** 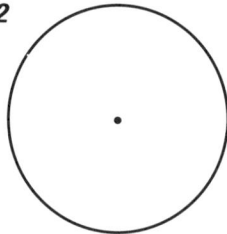   **13**   **14** 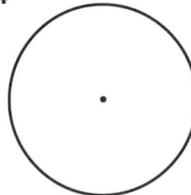   **15**

EXERCISE **12.5**

Write your answers correct to 3 significant figures.

Find the area of circles with radius:

   1 3 cm   **2** 8 cm   **3** 7.5 cm   **4** 10.5 cm   **5** 45 mm

Find the area of circles with diameter:

   6 7 cm   **7** 10 cm   **8** 15 cm   **9** 40 cm   **10** 65 cm

11 What is the area of a garden pond of diameter 1.3 m?

12 The diameter of a dinner plate is 25 cm. What is its area?

13 A circular cake tin has a diameter of 15 cm. What is the area of the base?

14 The long hand of a clock measures 9 cm in length. What is the area of its clock face?

15 A garden sprinkler has a jet 4.5 metres long. What area of lawn can it keep watered as it rotates?

Investigation C

Find the area of the coins we use.
Are there any for which it is difficult to find the area. Why is this?

Investigation D

The diagram shows three circles arranged as closely together as possible so that the rectangle drawn around them is of the smallest area. Could you arrange three circles in a different way to make the rectangle drawn around them have an even smaller area?

   Try the same investigation with four circles: arrange them in such a way that the rectangle drawn around them has the smallest possible area. Repeat with five, six circles, etc.

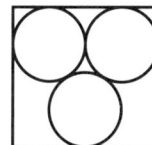

Example 5

Find the area of the shaded region.

The area of the shaded region is

area of larger circle − area of smaller circle

$$= \pi \times 3 \times 3 - \pi \times 2 \times 2$$
$$= 28.274 - 12.566 = 15.708 \text{ cm}^2$$

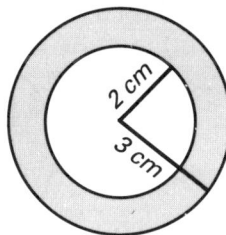

Example 6

Find the area of the shaded section.

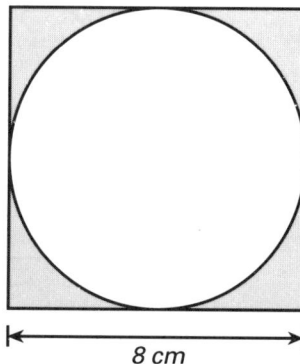

The area of the shaded section is

area of square − area of circle
$$= 8 \times 8 - \pi \times 4 \times 4$$
$$= 64 - 50.265 = 13.735 \text{ cm}^2$$

8 cm

EXERCISE 12.6

1 Find (*a*) the area of the larger circle (*b*) the area of the smaller circle
(*c*) the area of the shaded region.

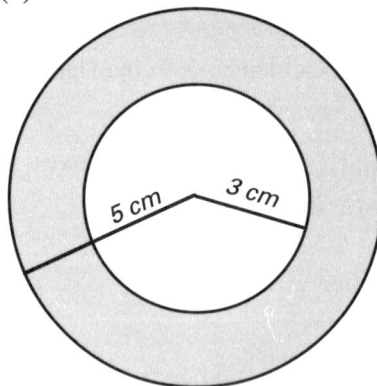

5 cm   3 cm

2 Find (*a*) the area of the rectangle (*b*) the area of the circle (*c*) the area
of the shaded region.

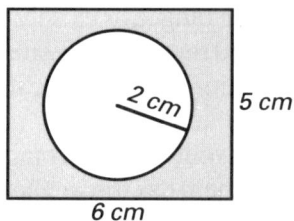

2 cm   5 cm

6 cm

3 Find (*a*) the area of the circle of radius 6 cm (*b*) the area of the
square of side 8 cm (*c*) the area of the shaded region.

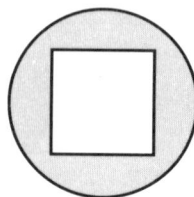

4 Find (*a*) the area of each circle of radius 3 cm (*b*) the area of the
rectangle (*c*) the area of the shaded region.

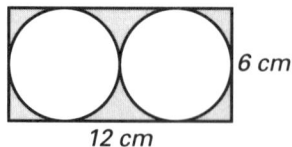

6 cm

12 cm

5 Find (*a*) the area of each circle of diameter 0.5 m (*b*) the area of the
rectangle (*c*) the area of the shaded region.

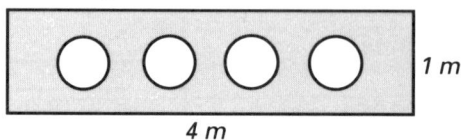

1 m

4 m

Combined shapes

Example 7

Find (a) the total area (b) the perimeter.

Shape A is a semicircle: area $= \pi r^2 \div 2 = \pi \times 2 \times 2 \div 2 = 6.28\,\text{m}^2$
Arc length of A $= \pi \times$ diameter $\div 2 = \pi \times 4 \div 2 = 6.28\,\text{m}$
Shape B is a rectangle of area $4 \times 6 = 24\,\text{m}^2$
Shape C is a semicircle of the same shape as A.

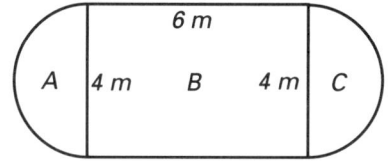

Total area $= 6.28\,\text{m}^2 + 24\,\text{m}^2 + 6.28\,\text{m}^2 = 36.56\,\text{m}^2 = 36.6\,\text{m}^2$ (to 3 s.f.)

Perimeter $= 6.28\,\text{m} + 6\,\text{m} + 6.28\,\text{m} + 6\,\text{m} = 24.56\,\text{m} =$
        $24.6\,\text{m}$ (to 3 s.f.)

EXERCISE 12.7

For each of the following shapes find (a) the total area (b) the perimeter.

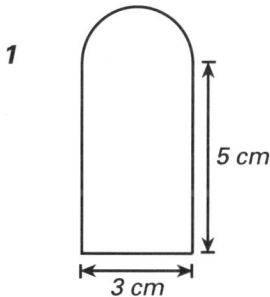

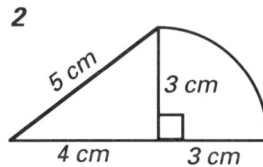

1

5 cm

3 cm

2

5 cm

3 cm

4 cm   3 cm

3 Find the area of the shaded region.

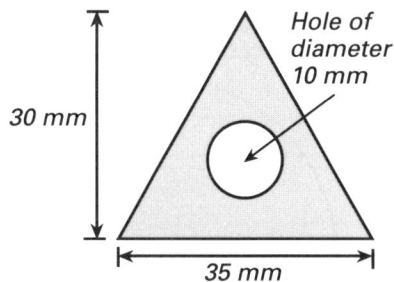

Hole of
diameter
10 mm

30 mm

35 mm

4 Find the area of the shaded region.

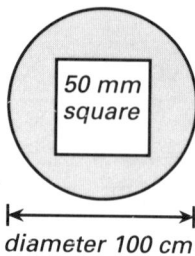

50 mm square

diameter 100 cm

5 The semicircle has a hole cut out of it. Find the area of the remaining piece.

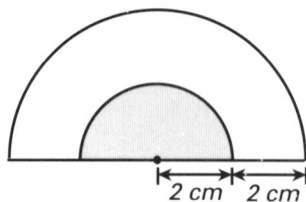

2 cm   2 cm

6 Find (*a*) the perimeter (*b*) the total area.

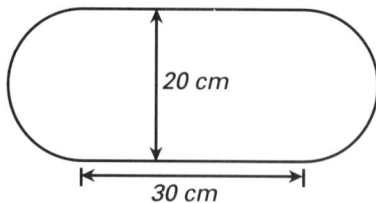

20 cm

30 cm

7 Find (*a*) the perimeter (*b*) the total area.

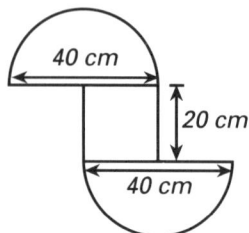

40 cm

20 cm

40 cm

8 Calculate the area once the triangular section has been removed.

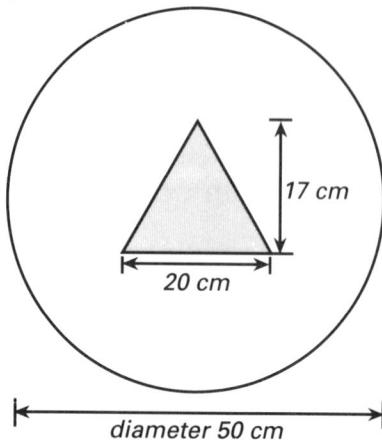

17 cm

20 cm

diameter 50 cm

9 This metal plate has had two holes drilled through it. Find the area of the plate.

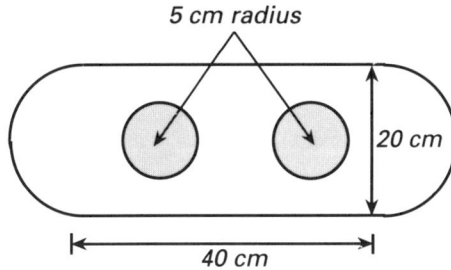

5 cm radius

20 cm

40 cm

Volume

The volume of a solid body is a measure of the space enclosed inside that solid. We already know how to find the volume of a cuboid.

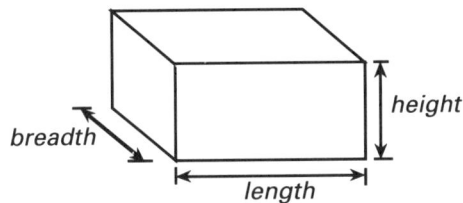

height

breadth

length

Volume = length × breadth × height

Frequently the volume of a solid is determined by the area of the end of the solid:

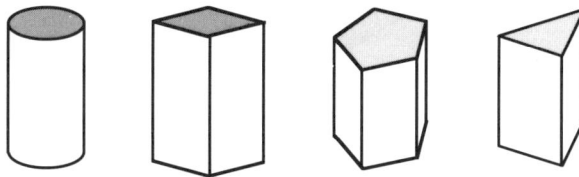

All these shapes have a **cross-section** which is shaded. These shapes have been built up from the cross-section to a certain height to give a solid body which has a volume.

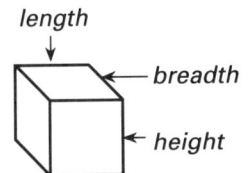

length

breadth

height

Example 8

The cuboid can be seen as a cross-section of area: length × breadth, and a height. The volume is therefore length × breadth × height, as we already know.

Example 9

The diagram represents a chocolate bar in the shape of a prism. The area of cross-section is 3.5 cm². Find the volume.

The volume is $3.5 \times 10 = 35$ cm³.

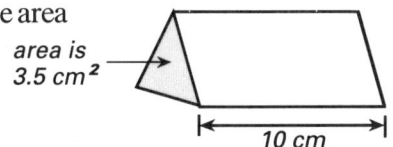

area is 3.5 cm²

10 cm

EXERCISE **12.8**

1 Find the volume of this triangular prism of cross-sectional area 8 cm².

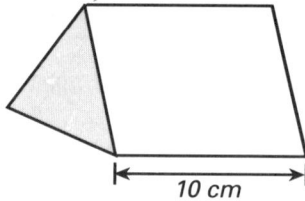

10 cm

2 Find the volume of this pentagonal prism of cross-sectional area 12.5 cm².

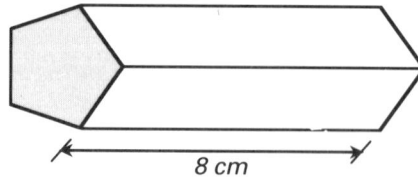

8 cm

3 Find the volume of this hexagonal prism of cross-sectional area 14 cm².

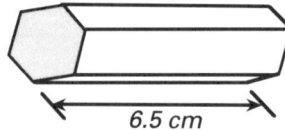

6.5 cm

4 Find the volume of this trapezoidal prism of cross-sectional area 35 cm².

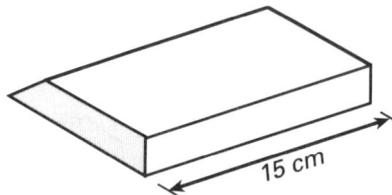

15 cm

5 Find the volume of this octagonal prism which is 20 cm deep.

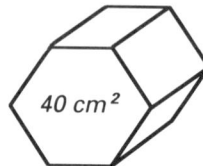

40 cm²

Find the volume of these prisms:

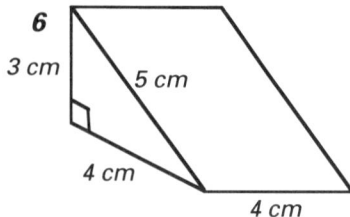

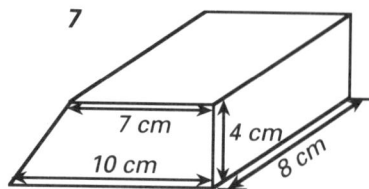

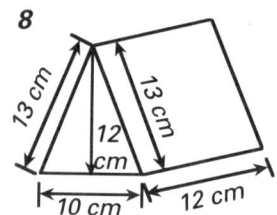

6
3 cm
5 cm
4 cm
4 cm

7
7 cm
10 cm
4 cm
8 cm

8
13 cm
12 cm
13 cm
10 cm
12 cm

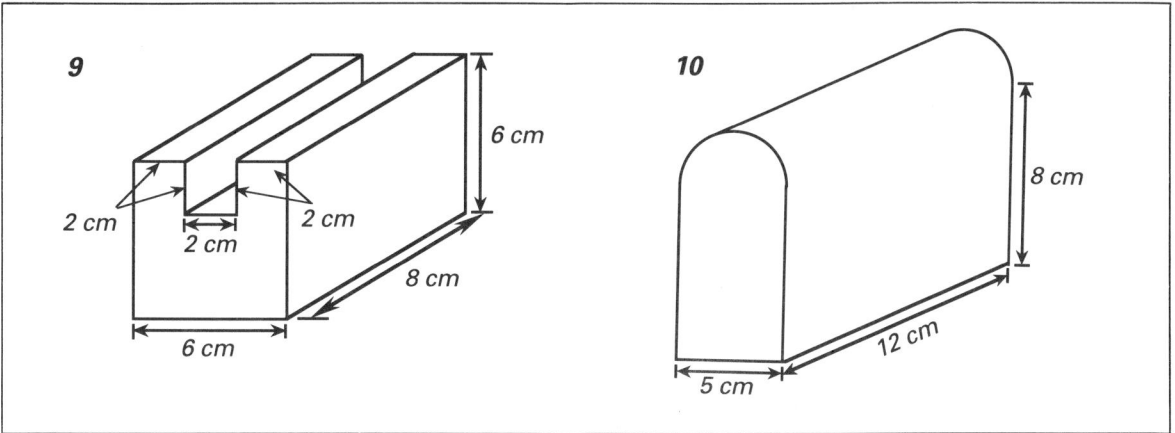

9

6 cm

2 cm   2 cm
2 cm

8 cm

6 cm

10

8 cm

12 cm

5 cm

The cylinder

Height

Radius

The cylinder is a special type of prism. The volume of a cylinder is still determined by the cross-sectional area × height.

The cross-sectional area is the area of its circular end $= \pi r^2$. So

volume of a cylinder $= \pi r^2 h$

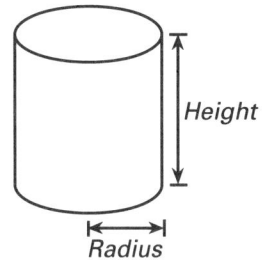

Example 10

Find the volume of this cylinder.

Volume $= \pi r^2 h = \pi \times 2 \times 2 \times 5 = 62.83 \, \text{cm}^3 = 62.8 \, \text{cm}^3$ (to 3 s.f.)

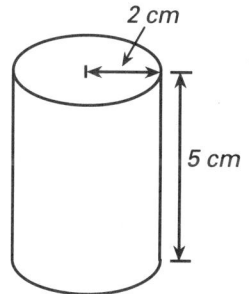

2 cm

5 cm

EXERCISE 12.9

Find the volume of these cylinders:

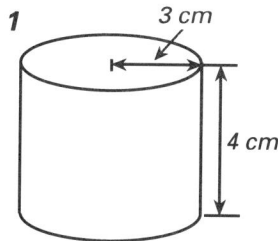

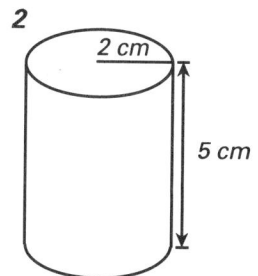

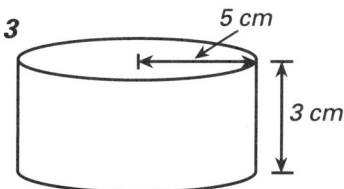

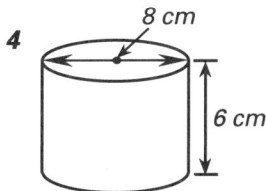

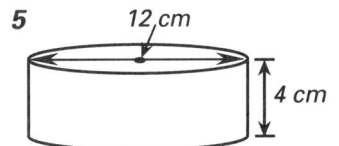

1   3 cm

4 cm

2   2 cm

5 cm

3   5 cm

3 cm

4   8 cm

6 cm

5   12 cm

4 cm

6 A cylinder with radius 2.5 cm and height 5 cm.

7 A tin of height 10 cm and diameter 5.5 cm.

8 A container of height 9 cm and radius 4.5 cm.

9 A jar of radius 4 cm and height 9 cm.

10 A bottle of height 15 cm and diameter 8.5 cm.

Investigation E

How many ways could you pack these cans into the box? What is the
maximum number of cans you can fit into the box? What is the minimum
volume of wasted space you would be left with in the box? A further
extension to this problem would be to find an actual tin or jar, and design
the net of a box to fit this tin, or any number of tins.

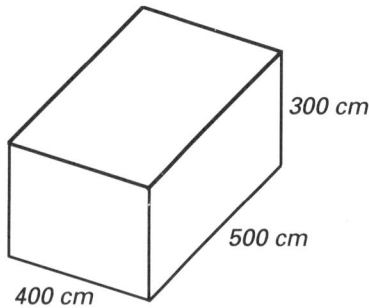

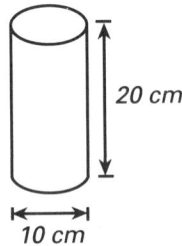

Investigation F

Repeat Investigation E, but this time consider the problem of packing
cuboids (packets) into the box, rather than cans.

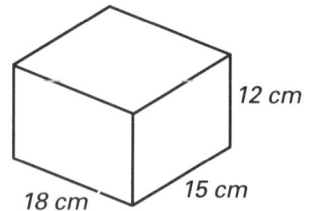

Problems with volume

Example 11

A two-litre bottle of lemonade is to be emptied into cylindrical cups of
radius 2 cm and 6 cm high. How many cups can be filled from the bottle?

1 litre $= 1000$ cm^3, so 2 litres $= 2000$ cm^3
Volume of a cup $= \pi r^2 h = \pi \times 2 \times 2 \times 6 = 75.398$ cm^3
Number of cups $= 2000 \div 75.398 = 26.5$ cups.

So the actual number of cups which can be filled is 26 cups: you must
always round down in these cases.

EXERCISE *12.10*

1 A three-litre bottle of pop is to be served in cylindrical paper cups of radius 2 cm and height 5 cm. How many cups can be filled from the bottle?

2 An ingot of metal 24 cm long, 5 cm wide and 4 cm high is to be melted down and recast into small medallions 0.25 cm thick and 2 cm in diameter. How many medallions can be made from the ingot?

3 A large consignment of tea is kept in a crate measuring 100 cm × 100 cm × 100 cm. It is to be divided into packets for sale, each packet measuring 4 cm × 4 cm × 6 cm. How many packets of tea can be filled from a crate?

4 Sand can be bought in containers as shown.

How many containers will need to be bought to fill a sand pit which measures 200 cm × 100 cm × 40 cm?

5 A large sack contains 50 000 cm^3 of sugar, and is to be repacked into boxes of sugar measuring 15 cm × 5 cm × 9 cm. How many boxes of sugar can be filled from each sack?

6 A large cylindrical drum has diameter 80 cm and height 150 cm. Its liquid contents are to be emptied into smaller containers each holding $1\frac{1}{2}$ litres. How many of the smaller containers could be filled from the drum?

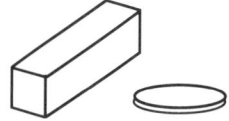

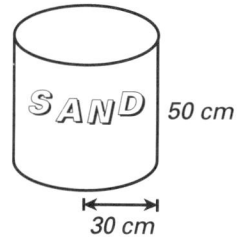

Edge, face and vertex

A **vertex** is a corner. A rectangular box has 8 vertices.
An **edge** is a straight line joining two vertices. A rectangular box has 12 edges.
A **face** is a surface, that is either a side, top or bottom of the box. A rectangular box has 6 faces.

The net of the cuboid shows more clearly the 6 faces, but cannot easily show the vertices. There are more than 12 edges in the net. Can you explain how they become 12 edges when the cuboid is assembled from the net?

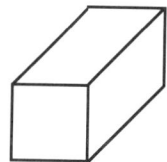

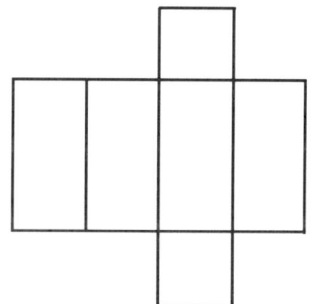

EXERCISE **12.11**

The following questions refer to nets and solids. The solids could be assembled from the nets if necessary.

1 This is the net of a triangular prism.

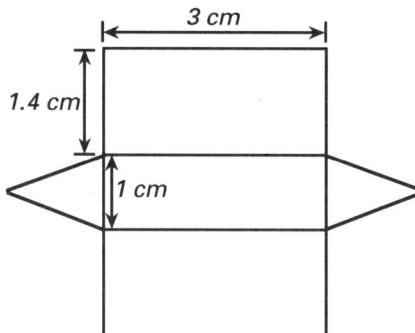

3 cm

1.4 cm

1 cm

    (*a*) For the prism, write down the number of (i) faces (ii) edges (iii) vertices.

    (*b*) What will be the volume of the prism?

    (*c*) Draw a different net for the same prism.

2 This is the net for a hexagonal prism.

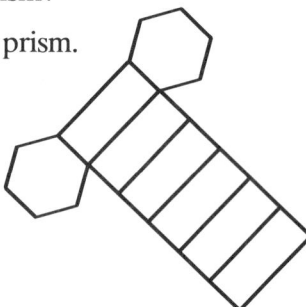

    (*a*) For the prism, write down the number of (i) faces (ii) edges (iii) vertices.

    (*b*) How many faces are rectangular?

    (*c*) Draw a different net for the same prism.

3 (*a*) Write down the number of (i) faces (ii) edges (iii) vertices.

    (*b*) What will be the volume of the prism?

    (*c*) Draw a net for the prism.

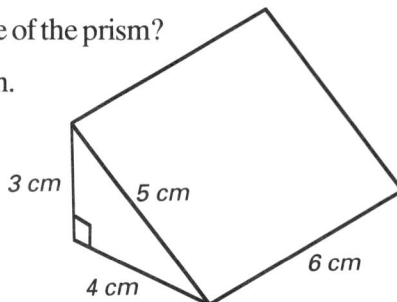

3 cm   5 cm

6 cm

4 cm

4 This is the net for a square pyramid.
For the prism, write down the number of (*a*) faces (*b*) edges
(*c*) vertices.

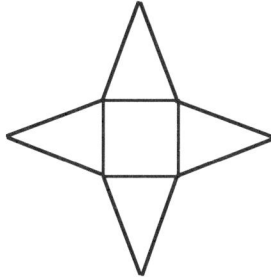

5 This is a triangular-based pyramid (a tetrahedron).

(*a*) Write down the number of (i) faces (ii) edges (iii) vertices.

(*b*) Draw a net for the tetrahedron.

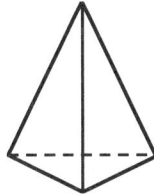

Investigation G

Solid	No. of faces	No. of vertices	No. of edges
Cuboid	6	8	12

Continue the table by listing the details of all the solids you have dealt
with in the previous questions. Can you describe a pattern in the numbers
in the table, and a relationship between the faces, vertices and edges?

13. Probability

As a reminder, let us look at some simple probability questions.

Example 1

When I throw a die, what is the probability of getting (*a*) a 4 (*b*) an even number (*c*) more than 4?

Probabilities are usually represented by fractions; the numerator (top number) is the number of ways that the answer we want can occur, and the denominator (bottom number) is the number of equally likely answers that could possibly occur.

(*a*) The probability of a 4 turning up is $\frac{1}{6}$, as there is only one 4 out of six equally likely outcomes.

(*b*) There are three even numbers possible. Hence

$$p(\text{even number}) = \frac{3}{6} \left(= \frac{1}{2} \right)$$

(*NB* p(even number) is shorthand for the probability of an even number occurring.)

(*c*) p(more than 4) = p(5 or 6) = $\frac{2}{6} \left(= \frac{1}{3} \right)$.

EXERCISE 13.1

1 I have five cards, the 6, 7, 8, 9 and 10 of clubs. If I pick one card at random, what is the probability that I pick (*a*) the 6 (*b*) an odd number (*c*) a number more than 7?

2 Three of my friends join me for a game of cards. The dealer of the first hand is chosen at random. What is the probability that I am *not* the one who is chosen from the four of us?

3 I put seven pieces of paper into a bag. On each piece is written a different day of the week. If I pick out one piece, what is the probability that the day written on it is (*a*) Tuesday (*b*) either Saturday or Sunday (*c*) not Wednesday?

4 There are five girls and three boys in our table tennis squad. One member of the squad is chosen at random to telephone the score to the local newspaper. What is the probability that the person chosen is (*a*) a boy (*b*) a girl?

5 A set of snooker balls consists of 15 reds, a white and 6 others. If one ball is packed at random, what is the probability that it is (*a*) red (*b*) white (*c*) one of the other colours?

6 A box contains five blue pens and three red pens. One pen is taken out without looking. What is the probability that it is red?

7 There have been 200 tickets sold in a raffle for a hamper, and I have bought 10 tickets. What is the probability that I win the hamper?

8 As I am shuffling a normal pack of 52 cards, one falls on to the table. What is the probability that it is (*a*) a king (*b*) a club (*c*) the four of spades?

9 On a stall at a school open day, 4 eggs and 46 eggshells are almost buried in sand, so that all you can see is what appears to be 50 eggs. If you pick one out, what is the probability that it will be (*a*) an eggshell (*b*) a good egg?

10 In a bag there are 10 buttons, identical apart from colour. Five are red, three are green and two are blue. One button is picked out at random. What is the probability that it is (*a*) red (*b*) *not* green (*c*) white (*d*) neither blue nor red (*e*) either red, green or blue?

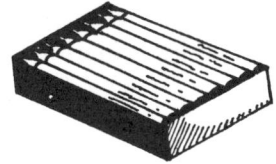

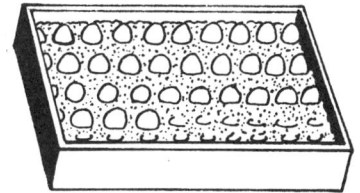

Question 10 in Exercise 13.1 reminds us that:

 (i) a probability of 0 means that the result is *impossible*;
(ii) a probability of 1 means that the result is *certain* to happen.

Experimental probability

If you throw a coin 20 times, and it lands heads 12 times and tails 8 times, would you say that the coin was biased? (Biased in this context means weighted to make heads more likely.)

   Probably not from this evidence – it is quite reasonable to expect a fair coin to come down heads 12 times out of 20. We may have doubts if heads occurred 16 or 17 times, especially if we repeated the 20 throws and the same thing happened.

   With coins, or dice, or a pack of cards, you can work out the probabilities without needing to perform any experiments, because we know that there are a certain number of *equally likely* possible outcomes (2 sides of a coin, 6 faces of a die, 52 cards in a pack). There are some events, however, where we cannot work out the probabilities exactly. We have to do some sort of experiment or trial in order to get a measure of the probabilities.

Example 2

When a drawing pin is dropped, what is the probability that it will land (*a*) on its head, like this: (*b*) with the point and edge resting on the

table, like this: ?

We cannot answer this question, because we cannot work out from symmetry or equally likely events what the probabilities are.

    We therefore have to base our answers on results obtained by performing a trial or experiment: in this case by repeatedly throwing the drawing pin and counting the number of time it lands like this   or like this .

    Naturally our results can only be estimates, but the more trials we perform the better the estimates can be expected to be.

Example 3

Out of 100 throws, a drawing pin landed on its head 70 times. Estimate the probability that the drawing pin lands on its head.

The probability of the drawing pin landing on its head is $= \frac{70}{100} \left(= \frac{7}{10} \right)$.

Also, the probability of it landing like this   is $\frac{30}{100} = \frac{3}{10}$.

Example 4

Our fourth year soccer team has scored 40 goals this season so far. Stan has scored 15 goals, and Lee and Geoff have each scored 10 goals. Based on these figures, what is the probability that our next goal will be scored by (*a*) Stan, (*b*) Geoff, (*c*) none of these three?

(*a*)   p(Stan scores) $= \frac{15}{40} \left(= \frac{3}{8} \right)$

(*b*)   p(Geoff scores) $= \frac{10}{40} \left(= \frac{1}{4} \right)$

(*c*)   These three scored $15 + 10 + 10 = 35$ goals.
    Other players scored the other 5 goals.
    So the probability that the next goal is scored by none of these three

    players (i.e. someone else) $= \frac{5}{40} \left(= \frac{1}{8} \right)$.

As in the drawing pin example, we can estimate the probabilities only on the goals actually scored. We could not have said at the start of the season what the probabilities are likely to be, except perhaps we could have a very rough idea from the performance of players when they were in the third year. There may have been changes in the team from the third year to the fourth year which could drastically alter the scoring potential of any of the players.

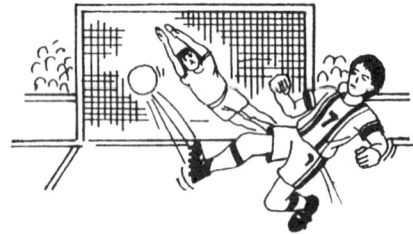

Another point to bear in mind is that these estimates of probability change with each new result. Initially this makes quite a difference, but after a large number of goals the change in probabilities is relatively small.

Example 5

Of the last 20 telephone calls that I have received, 4 have been 'wrong numbers'. If the telephone rings now, what is the probability, based on these figures, that it is a wrong number?

The probability that it is a wrong number is $\frac{4}{20}(=\frac{1}{5})$.

Let us assume that the call was in fact a wrong number. What is the probability that the *next* call is a wrong number?

I have now had 21 calls, 16 being all right, and 5 being wrong numbers.

The estimated probability of a wrong number now is $\frac{5}{21}$.

Investigation A

Which fraction is larger (and therefore represents the greater probability), $\frac{4}{20}$ or $\frac{5}{21}$?

Investigate methods of comparing fractions, using (*a*) decimals (*b*) fractions.

EXERCISE 13.2

1 I have tipped a box of 150 identical drawing pins on to the table, and 90 have landed on their heads, like this ⚲. What is the

   probability that one of these drawing pins lands (*a*) like this ⚲ (*b*) like this ⚲ ?

2 The school bus has been late four times out of a possible fifteen in the last three weeks. Estimate the probability that it will *not* be late today.

3 In testing car windscreens, a weight is dropped from a fixed height on to each windscreen. Out of 60 windscreens tested, 7 shattered into little pieces. What is the probability that the next one tested will shatter?

4 In his last ten frames of snooker, Davis Steve scored 57, 38, 6, 91, 62, 41, 59, 17, 47, and 46 points. Based on these scores, what is the probability that he scores over 50 in his next frame?

5 Our cricket team has scored over 150 runs on 15 occasions so far this season. In 11 of these 15 games, someone on our team scored over 50 runs.

   If we score 150 in our next game, what is the probability that someone on our team will score over 50 runs?

Investigation B

(a) Shuffle a pack of cards and turn over the top card.

(b) If it is a *heart*, enter a tally mark in the '1' row of the table and start again.

(c) If it is not a heart, turn over the next card.

(d) If this card is a heart, enter a tally mark in the '2' row, and start again.

(e) Continue turning over the cards, one by one, until you get a heart, entering a tally mark in the appropriate row.

(f) Repeat the experiment, until you have 20 results in your table.

No. of cards turned over until a heart turns up	Tally	Total
1		
2		
3		
4		
5		
6		
7		
8		
.		

On this evidence, what is the probability that the first heart turns up on the *third* card turned over?

Combined probabilities

Example 6

What is the probability that a coin, when tossed twice, lands 'heads' both times?

There are two possible ways of tackling this problem:
(i) by considering all possible outcomes,
(ii) by considering the probability of each step in turn.

(i) If we let H mean 'lands heads' and T means 'lands tails', then there are four equally likely outcomes:

   H and H,   H and T,   T and H   or   T and T

So the probability of two heads is $\frac{1}{4}$, as there is only one way of getting two heads, out of four equally likely results.

(ii) On the first throw, the probability of a head is $\frac{1}{2}$. Similarly, the probability of a head on the second throw is also $\frac{1}{2}$: to get the second head, we need $\frac{1}{2}$ of the probability of getting the first head. The probability we need is therefore $\frac{1}{2}$ of $\frac{1}{2} = \frac{1}{4}$.

Example 7

Two dice are thrown. What is the probability of scoring a total of 11? (Throwing two fair dice is the same as throwing one of them twice, as far as calculating probabilities goes.)

(i) Although there are eleven possible totals (2, 3, 4, 5, 6, 7, 8, 9, 10, 11 and 12) they are not all equally likely.

   As each die can turn up in six ways, there are 36 possible equally likely outcomes for two dice. They can be listed systematically like this:

$$
\begin{array}{cccccc}
(1,1) & (1,2) & (1,3) & (1,4) & (1,5) & (1,6) \\
(2,1) & (2,2) & (2,3) & (2,4) & (2,5) & (2,6) \\
(3,1) & (3,2) & (3,3) & (3,4) & (3,5) & (3,6) \\
(4,1) & (4,2) & (4,3) & (4,4) & (4,5) & (4,6) \\
(5,1) & (5,2) & (5,3) & (5,4) & (5,5) & (5,6) \\
(6,1) & (6,2) & (6,3) & (6,4) & (6,5) & (6,6)
\end{array}
$$

As there are two ways of scoring 11, i.e. (5,6) and (6,5), the probability of scoring 11 is $\frac{2}{36} (= \frac{1}{18})$.

(ii) To score 11, you need either a 5 followed by a 6 or a 6 followed by a 5.

   Each of these outcomes has a probability of $\frac{1}{6} \times \frac{1}{6} = \frac{1}{36}$.

So together the probability of scoring 11 is $\frac{1}{36} + \frac{1}{36} = \frac{2}{36} (= \frac{1}{18})$.

Investigation C

Find the probability of scoring each of the eleven possible totals when two dice are thrown.

EXERCISE **13.3**

1 I have two blue pens and one red pen in my pocket. After picking one pen at random, and writing with it, I return it to my pocket. Later I once more select a pen at random.

(*a*) What is the probability that the first pen is blue?

(*b*) Work out the probability that both pens are blue.

2 I throw a coin and a die. I win if I throw a head, and either a 3 or a 4.

(*a*) List all the possible outcomes e.g. (H,1), (H,2), ..., (T,6).

(*b*) What is the probability that I win?

3 As an experiment I shuffle four cards, the jack, queen, king and ace of spades, pick one, and make a note of it. I then shuffle all four cards and pick one again.
   By listing all possible pairs of outcomes [(J,J,), (J,Q), ..., (A,A)] or taking each selection in turn, work out the probability of picking (*a*) two aces (*b*) a king and a queen (*c*) any 'pair'.

4 I had a 10p, a 5p and a 2p piece in a pocket which also had a handkerchief in it. When I took the handkerchief out, a coin fell on to the floor. I put the coin and the handkerchief back in my pocket. Later I took the handkerchief out, and again a coin fell out. (I have now put the handkerchief in another pocket!)
   By considering each event in turn, or by listing all possible outcomes, calculate the probability that

(*a*) the 2p coin fell out twice,

(*b*) the 5p coin did *not* fall out on either occasion.

5 I have three counters. On one of them I write the number '3', on another '4' and on the third '5'. The counters are put into a bag, shaken, and one is picked out. After noting the number, it is returned to the bag, which is shaken and another is picked out. The number on this counter is added to the first number.

(*a*) List all the possible ways in which the numbers could come out.

(*b*) What is the probability that the total is (i) 10 (ii) 6 (iii) 7 (iv) 11?

Selection without replacement

In question 3 of Exercise 13.3, two random selections were made from a set of four cards. The probability of picking an ace was $\frac{1}{4}$ in each case.

If, however, you keep your first card, and then make your second selection from the remaining three cards, your answers are going to be different from the example above where the card was replaced before the second selection was made.

Indeed, if you keep your first card, it is impossible to select a pair of aces!

Example 8

List all the possible pairs you could select from the four cards jack, queen, king and ace of spades.

Calculate the probability that in the pair you select you have (*a*) the ace (*b*) both the king and the queen (*c*) *not* the jack.

There are six possible pairs of cards:

  A,K   A,Q   A,J   K,Q   K,J   Q,J

(*a*) Three of these contain the ace.
   The probability that the pair will contain the ace is $\frac{3}{6}(=\frac{1}{2})$.
(*b*) Only one pair has the king and queen.
   Therefore $p(K,Q)=\frac{1}{6}$.
(*c*) Three of the pairs do not have the jack: A,K; A,Q and K,Q. Therefore
   $p(\text{no jack})=\frac{3}{6}(=\frac{1}{2})$.

Example 9

I have four grey socks and six black socks in a drawer. If I take out one sock without looking, then another, what is the probability that I have taken out a pair of grey socks?

The probability that the first sock is grey is $\frac{4}{10}$. There are nine socks left in the drawer, three grey and six black. The probability that the second sock is grey is $\frac{3}{9}$.

The probability of taking out a pair of grey socks is therefore $\frac{3}{9}$ or $\frac{1}{3}$ of the $\frac{4}{10}$ probability that the first sock is grey.

This is $\frac{3}{9}\times\frac{4}{10}=\frac{12}{90}(=\frac{2}{15})$.

In general, for a repeated event (i.e. picking a sock) the probability of any outcome is worked out by *multiplying* the probabilities of the two separate outcomes.

The probability of two black socks would be $\frac{6}{10} \times \frac{5}{9} = \frac{30}{90}$ or $\frac{1}{3}$.

(There would be five black socks left, out of nine, once one black sock had been removed.)

Example 10

The four aces from a pack of cards are placed face down on a table. I pick one card, then another. What is the probability that I have picked up (*a*) the two black aces (*b*) the ace of hearts and the ace of spades (*c*) *not* the ace of diamonds?

Method 1. List all possible outcomes: CD CH CS DH DS HS.

(*a*) p(two blacks) = p(CS) = $\frac{1}{6}$.

(*b*) p(HS) = $\frac{1}{6}$.

(*c*) p(not a diamond) = p(CH or CS or HS) = $\frac{3}{6} (= \frac{1}{2})$.

Method 2

(*a*) 1st pick: p(black) = $\frac{2}{4}$.

    2nd pick: p(black) = $\frac{1}{3}$.

    Hence p(two blacks) = $\frac{2}{4} \times \frac{1}{3} = \frac{2}{12} = \frac{1}{6}$.

(*b*) 1st pick: p(heart) = $\frac{1}{4}$.

    2nd pick: p(spade) = $\frac{1}{3}$.

    Hence p(H then S) = $\frac{1}{4} \times \frac{1}{3} = \frac{1}{12}$.

    1st pick: p(spade) = $\frac{1}{4}$.

    2nd pick: p(heart) = $\frac{1}{3}$.

    Hence p(S then H) = $\frac{1}{4} \times \frac{1}{3} = \frac{1}{12}$.

    Hence p(heart and spade) = $\frac{1}{12} + \frac{1}{12} = \frac{2}{12} = \frac{1}{6}$.

(*c*) 1st pick: p(not diamond) = $\frac{3}{4}$.

    2nd pick: p(not diamond) = $\frac{2}{3}$.

    Hence p(neither card a diamond) = $\frac{3}{4} \times \frac{2}{3} = \frac{6}{12} = \frac{1}{2}$.

With a large number of possible outcomes, method 2 is often preferable.

Example 11

What is the probability of receiving a pair of sevens when you are dealt two cards?

There are $52 \times 51 = 2652$ possible outcomes – it would take quite a while to list them all!

$$p(\text{1st card is a 7}) = \frac{4}{52}.$$

$$p(\text{2nd card is a 7}) = \frac{3}{51}$$

as there are now only three sevens left out of 51 cards that could be dealt to you.

Hence $p(\text{7 and 7}) = \frac{4}{52} \times \frac{3}{51} = \frac{12}{2652} = \frac{1}{221}.$

EXERCISE 13.4

Work out these probabilities, using either of the two methods.

1 I have three blue pens and two red pens in my bag. If I pick out two pens, one after the other, find the probability that I pick (*a*) two blue (*b*) two red (*c*) one of each (*d*) no blue.

2 My gran has £1 to give to my sister and myself. She changes it for a 50p, two 20p pieces and a 10p piece, and gives us two coins each, at random. What is the probability that I get (*a*) 40p (*b*) 70p (*c*) 50p?

3 Of the six chocolates remaining in a box, there are two milk ones and four plain. The six chocolates are covered in identical wrappers. If I pick one chocolate, then another, what is the probability that I pick the two milk chocolates?

4 The names of the nine finalists in a competition are drawn at random. Of these finalists, six are women and three are men. What is the probability that the first two names drawn are both women?

5 For a particular card game, only the 2, 3, 4, 5 and 6 of spades are used. I am dealt one of these cards, followed by another. What is the probability that my two cards (*a*) are the 3 and 4 (*b*) add up to 8 (*c*) contain the 6?

6 Two students are selected at random from a group of five boys and two girls. What is the probability that they are (*a*) both boys (*b*) both girls?

7 Our youth group is holding a raffle. Four hundred tickets have been sold, and I have bought five. What is the probability that the number on one of my tickets is the *second* number drawn out?

14. Trigonometry

Trigonometry is concerned with the study of right-angled triangles. Land surveyors, ocean navigators and astronomers were among the first to study this subject.

Right-angled triangles

A right-angled triangle is simply a triangle in which there is a right angle.
The sides of a right-angled triangle are given special names. The side opposite the right angle, the longest side, is known as the **hypotenuse**. You came across the hypotenuse in Chapter 5 when you were using Pythagoras' theorem. Side BC in the diagram is opposite the angle x, and is called the **opposite** side. The side AC next to the angle is called the **adjacent** side.

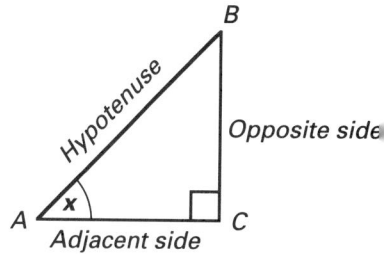

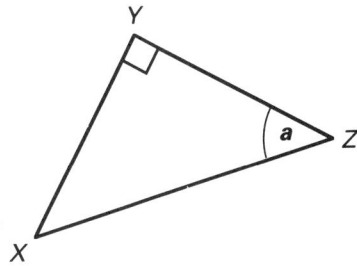

Example 1

In this triangle XZ is the hypotenuse, XY is the opposite side because it is opposite the angle a, so YZ is the adjacent side.

EXERCISE 14.1

Copy each triangle, and write in the names of each of the sides.

1

2

3

4

5

6

7

8

9

Ratio of sides

In this triangle the hypotenuse measures 5 m, and the side opposite angle
x measures 4 m. We say the **ratio** of

$$\frac{\text{opposite side}}{\text{hypotenuse}} = \frac{4}{5}$$

There are two other combinations of these sides we could use to write
down ratios:

$$\frac{\text{adjacent side}}{\text{hypotenuse}} = \frac{3}{5}$$

$$\frac{\text{opposite side}}{\text{adjacent side}} = \frac{4}{3}$$

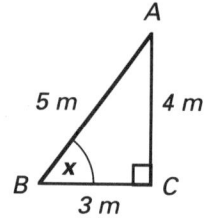

EXERCISE 14.2

Write down the three ratios possible for each of the following
right-angled triangles:

1

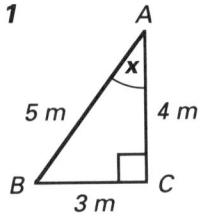

2

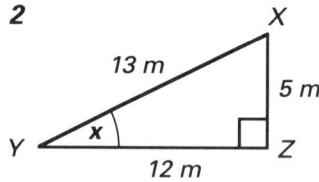

3

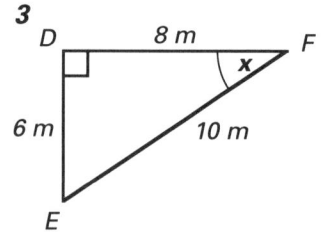

4

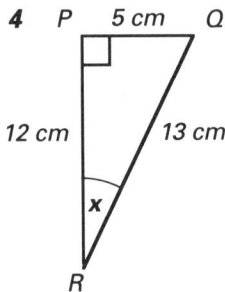

5

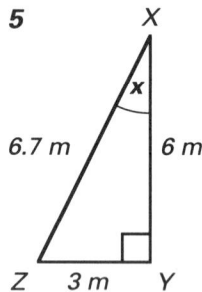

6

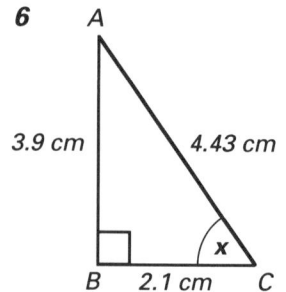

On a sheet of graph paper draw accurately the triangle as shown. By finding the actual lengths of the sides BC and AB, write down the ratio of the sides:

$$\frac{\text{opposite side}}{\text{hypotenuse}} = \frac{BC}{AB}$$

Change this fraction into a decimal by the division: BC ÷ AB.

Now draw a different right-angled triangle, but still with the same angle of 35°. Work out the ratio and find the decimal equivalent. You should get the same answer as before. Do you think you will always get the same answer? Draw a third right-angled triangle with a 35° angle and repeat the exercise.

All the triangles you have drawn are *similar triangles*, so the ratio of any two corresponding sides will never change and you should always get an answer of about 0.57 using an angle of 35°. Will we have the same answer if we change the angle?

Draw another triangle with an angle of 40° and find the decimal answer for the same ratio. This time you should have an answer of about 0.64, so changing the angle also changes the decimal answer.

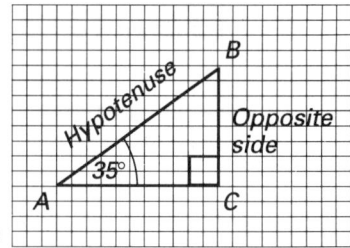

Sine

The ratio we have been using, $\frac{\text{opposite side}}{\text{hypotenuse}}$, is called the **sine** ratio.

We usually shorten the word sine to sin, and we can talk about the sine of the angle 35° as being 0.57, or sin 35° = 0.57 and sin 40° = 0.64.

EXERCISE 14.3

Using the same method as described above, draw triangles on graph paper, and by measurement find the sine ratio for each of the following angles.

1 45°   **2** 60°   **3** 30°   **4** 50°   **5** 70°   **6** 20°   **7** 65°   **8** 55°

Using calculators

It would be tedious to have to draw a triangle on graph paper every time we wanted to know the sine of an angle. Today many calculators are equipped to do this for us. For the work in this chapter you need a calculator with a button, or function, labelled SIN, which is the same sine ratio we have been finding by drawing triangles.

To find the sine of the angle 35°, first enter 35, then press the $\boxed{\text{SIN}}$ button. The answer you should get is 0.5735764, which is more accurate than we found before by drawing triangles. We normally write answers to four decimal places, as we did in Chapter 5. So, sin 35° = 0.5736.

EXERCISE 14.4

Using your calculator, write down the sines of the following angles to four decimal places.

1 45° *2* 60° *3* 62° *4* 75° *5* 43° *6* 39° *7* 3° *8* 21°

9 89° *10* 31° *11* 39° *12* 17° *13* 73° *14* 44° *15* 69° *16* 71°

17 88° *18* 5° *19* 0° *20* 90° *21* 45.3° *22* 62.4° *23* 40.3° *24* 10.1°

25 77.2° *26* 87.5° *27* 75.4° *28* 21.9° *29* 12.7° *30* 47.3° *31* 13.3° *32* 60.2°

33 28.3° *34* 18.8° *35* 7.7°

Using sine

We can now use what we know about the sine ratio in a right-angled triangle to help us find the opposite side in the triangle, if we know an angle and the hypotenuse.

Example 2

We will use the letter x for the side we want to find.

$$\frac{\text{opposite side}}{\text{hypotenuse}} = \frac{AB}{AC} = \frac{x}{8}$$

$$\sin 38° = \frac{x}{8}$$

$$0.6157 = \frac{x}{8}$$

$$x = 0.6157 \times 8 = 4.9256 = 4.93 \text{ cm (to 3 s.f.)}$$

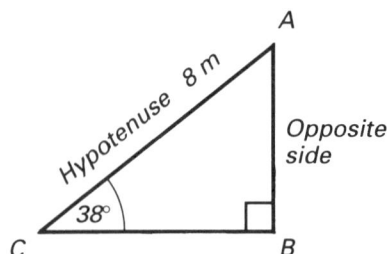

EXERCISE **14.5**

Find the side marked x in each of these triangles, writing your answers to 3 s.f.

1
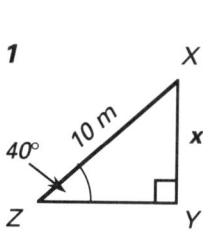

2
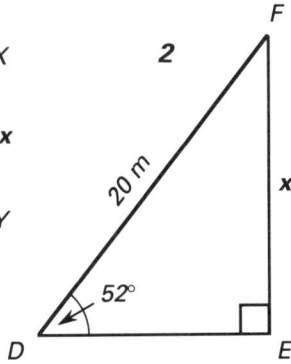

3
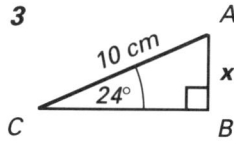

4
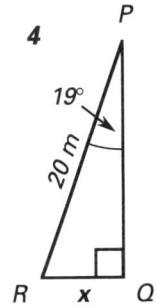

5
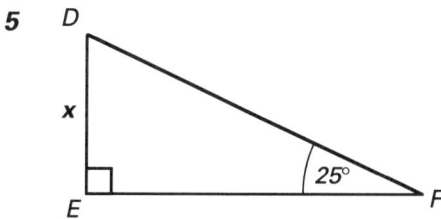

6
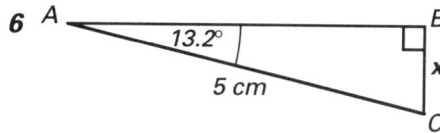

7
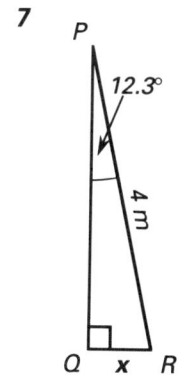

8
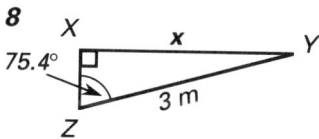

9
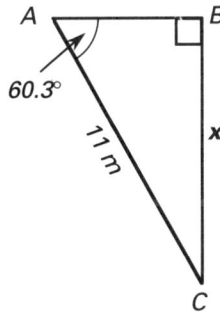

10
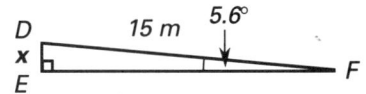

EXERCISE **14.6**

Find the side marked x in each of these triangles, writing your answers to 3 s.f.

1
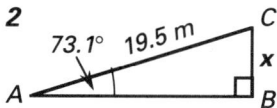

2
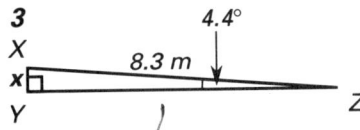

3
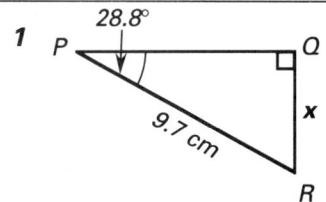

4
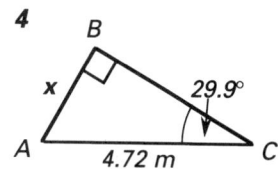

5 37.4°
P   Q
5.8 cm
R
x

6 D
48.5°
12.6 cm
E   x   F

7 S   x   T
70.1°
15.5 cm
U

8 A   42.6°
5.5 cm
B   x   C

9
Q
40.9°
x
P   9.75 m   R

10 Y   x   Z
60.3°
15.9 cm
X

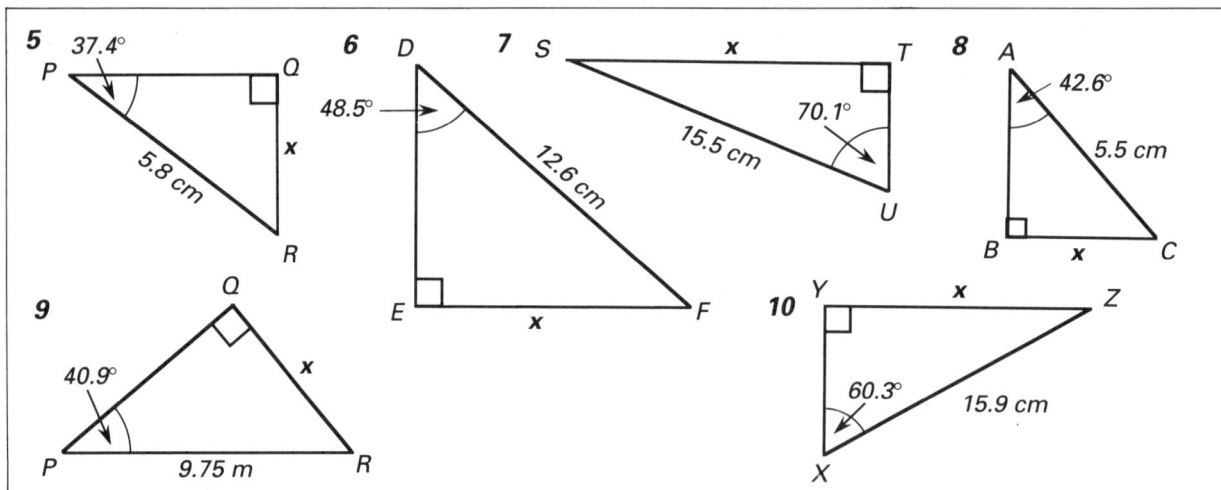

Finding the angle

On some occasions we are given two sides of the triangle and we need to find one of the missing angles. Again we can use trigonometry to help us.

Example 3

If the sine of an angle is 0.91, what is that angle?

On a calculator enter the decimal 0.91.
   The calculator can now be used to work out the angle. Most calculators require you to press two buttons: firstly ARC , 2ND FCT , or INV followed by SIN . So you should press:

     ARC SIN ; 2ND FCT SIN ; INV SIN , or an appropriate alternative.
     (Some calculators have a *single* button which will do this for you, such as SIN⁻¹

This should give you an answer of 65.505352°, or 65.5° to one decimal place, as we normally write angles in degrees to one decimal place.

EXERCISE 14.7

Find the angle, to one decimal place, whose sine is:

1 0.5   **2** 0.72   **3** 0.57   **4** 0.62   **5** 0.25   **6** 0.43

7 0.89   **8** 0.41   **9** 0.97   **10** 0.432 **11** 0.597 **12** 0.653

13 0.902 **14** 0.908 **15** 0.482 **16** 0.571 **17** 0.863 **18** 0.713

19 0.998 **20** 0.002

Example 4

We will use the letter x for the angle we want to find.

$$\frac{\text{opposite side}}{\text{hypotenuse}} = \frac{\text{AB}}{\text{AC}} = \frac{3.5}{9}$$

$$\sin x = \frac{3.5}{9}$$

$$\sin x = 0.3888$$

$$x = 22.9° \text{ (to 1 decimal place)}$$

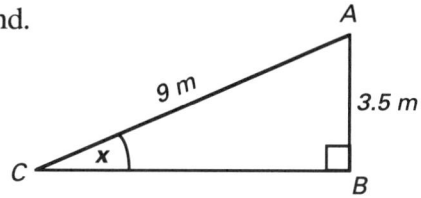

EXERCISE **14.8**

Find the angle marked x in each triangle, writing your answers to one decimal place.

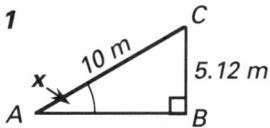

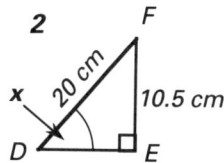

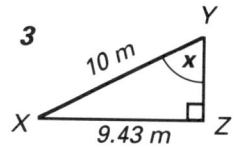

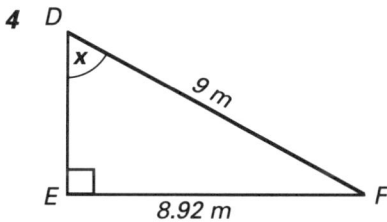

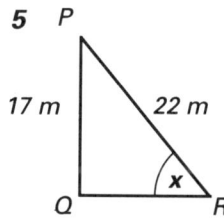

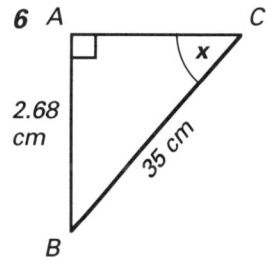

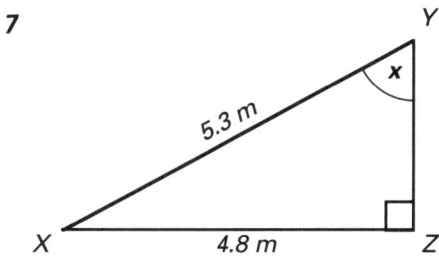

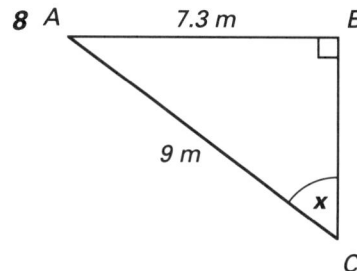

1

2

3

4

5

6

7

8

EXERCISE 14.9

Find the angle marked *x* in each triangle, writing your answers to
one decimal place.

1
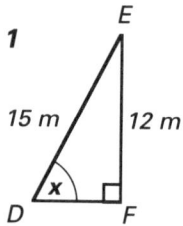

2
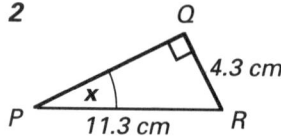

3
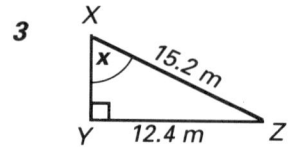

4
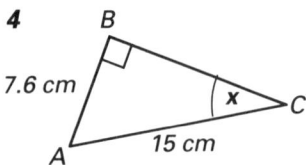

5
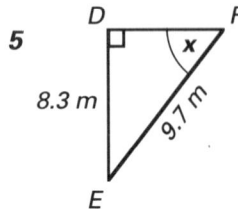

6
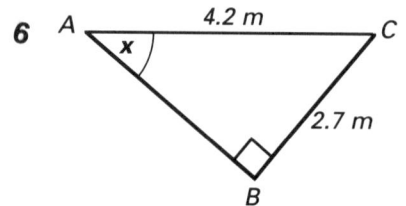

7
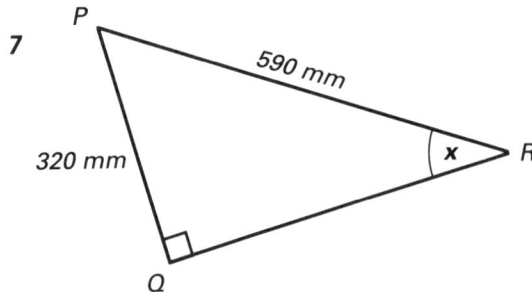

Cosine

The **cosine** of an angle is the ratio $\dfrac{\text{adjacent side}}{\text{hypotenuse}}$.

For this triangle we could write:

$$\cos x = \frac{\text{adjacent side}}{\text{hypotenuse}} = \frac{3}{5} = 0.6$$

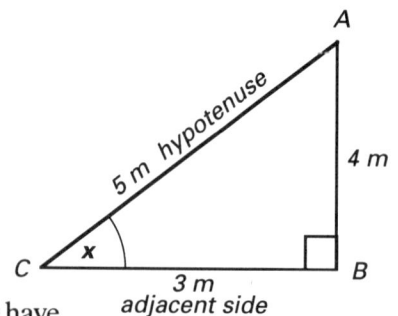

We need cosines as well as sines since the cosine helps us when we have
only an adjacent side and a hypotenuse.

   As with sines, the calculator can help us find the cosine of any angle as
a decimal ratio. We can use the $\boxed{\text{COS}}$ button on the calculator. If instead
we knew the decimal and wanted to find the angle, we would work
backwards as before, using $\boxed{\text{INV}}$ $\boxed{\text{COS}}$, $\boxed{\text{2ND FCT}}$ $\boxed{\text{COS}}$, $\boxed{\text{ARC}}$
$\boxed{\text{COS}}$, $\boxed{\text{COS}^{-1}}$, etc.

Example 5

cos 25.6° = 0.902 (to 3 decimal places).
The angle whose cosine is 0.54 is ... 57.3° (to 1 decimal place).

EXERCISE 14.10

Find the cosine of the following angles, writing your answers to four decimal places.

1 32°   **2** 67°   **3** 72°   **4** 12°   **5** 29°   **6** 51.8°

7 45.5°   **8** 21.2°   **9** 75.4°   **10** 43.6°   **11** 3.1°   **12** 12.2°

13 87.9°   **14** 39.4° **15** 30.7° **16** 87.4°   **17** 45.1° **18** 24.7°

19 49.9°   **20** 5.1°   **21** 78.5° **22** 21.3°   **23** 77.8° **24** 64.9°

Find the angle, to one decimal place, whose cosine is:

25 0.8829   **26** 0.2588   **27** 0.9715   **28** 0.3518   **29** 0.8100

30 0.1716   **31** 0.4982   **32** 0.9633   **33** 0.4040   **34** 0.8009

35 0.1705   **36** 0.1703   **37** 0.3353   **38** 0.9518   **39** 0.1728

40 0.4825   **41** 0.9274   **42** 0.0741

Using cosine

Example 6

We will use the letter x for the side we want to find.

$$\frac{\text{adjacent side}}{\text{hypotenuse}} = \frac{DE}{DF} = \frac{x}{5}$$

$$\cos 30° = \frac{x}{5}$$

$$0.8660 = \frac{x}{5}$$

$$x = 0.8660 \times 5 = 4.33 \text{ cm}$$

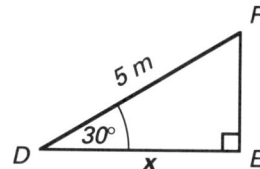

Example 7

We will use the letter x for the angle we want to find.

$$\frac{\text{adjacent side}}{\text{hypotenuse}} = \frac{PQ}{PR} = \frac{5.5}{10}$$

$$\cos x = \frac{5.5}{10}$$

$$\cos x = 0.55$$

$$x = 56.6° \text{ (to 1 decimal place)}$$

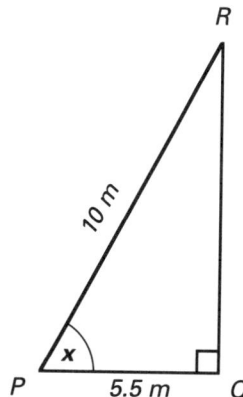

EXERCISE **14.11**

Calculate the side marked x in each of the triangles, writing your
answers to 3 s.f.

1
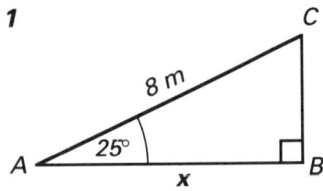

2
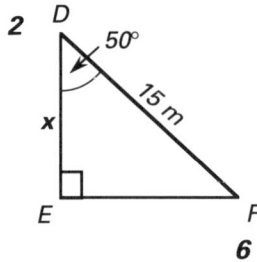

3
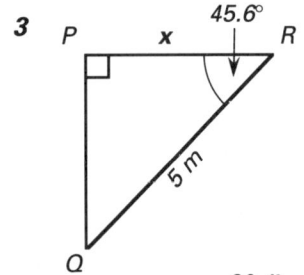

4
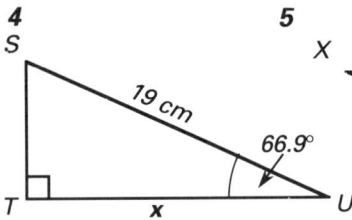

5
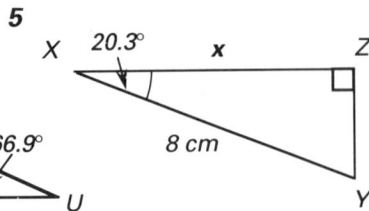

6
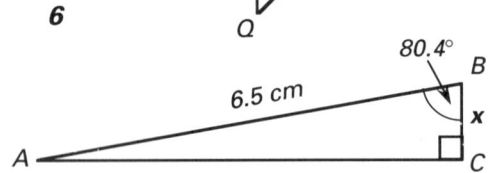

7
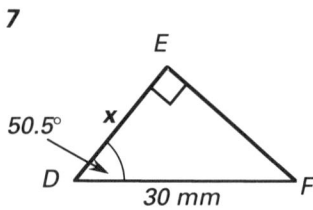

8
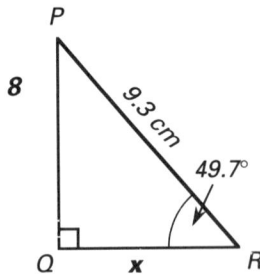

9
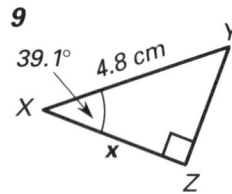

10
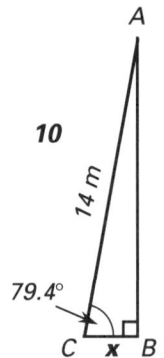

EXERCISE **14.12**

Calculate the angle marked x in each of the triangles, writing your
answers to one decimal place.

1
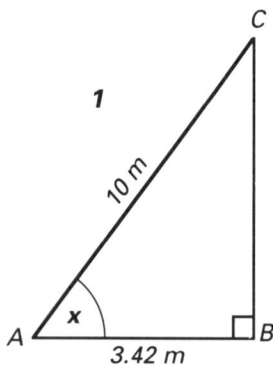

2
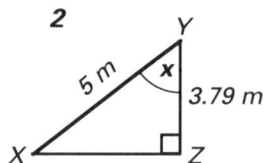

3
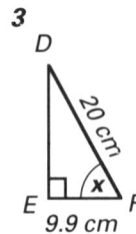

4
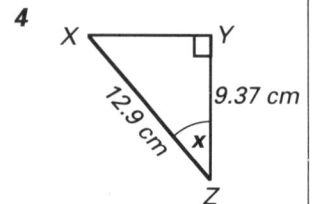

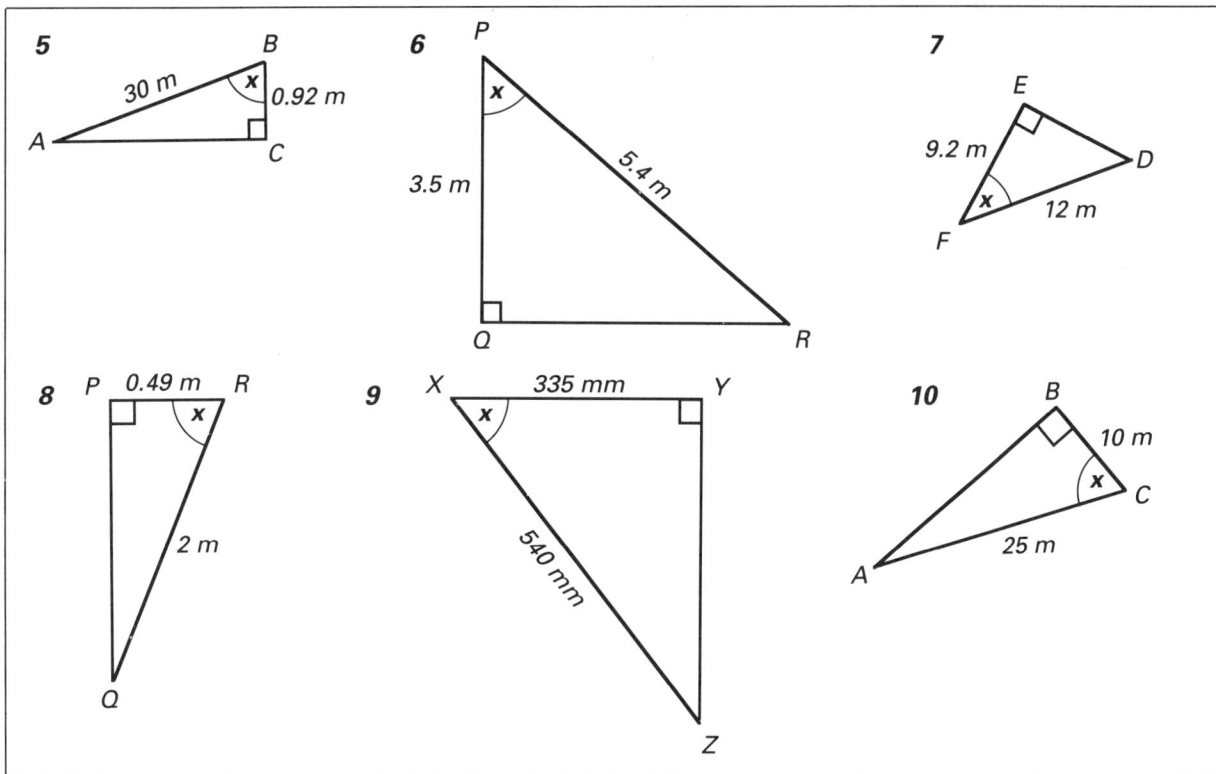

Investigation A

Investigate the relationship between sin and cos. Pick several angles of which to find both the sin and cos, writing down your results. Do you notice anything about the results? Draw up a table to record your results, and, if necessary, pick a complete range of angles to try. Write down clearly what you find.

Tangent

The **tangent** of an angle is the ratio $\dfrac{\text{opposite side}}{\text{adjacent side}}$.

For this triangle we could write:

$$\tan x = \frac{\text{opposite side}}{\text{adjacent side}} = \frac{4}{3} = 1.3333$$

We need tangents as well as sines and cosines since the tangent helps us when we only have an adjacent and an opposite side, that is, no hypotenuse.

As with sines and cosines, the calculator can help us find the tangent of any angle as a decimal ratio. We can use the $\boxed{\text{TAN}}$ button on the calculator. If instead we knew the decimal and wanted to find the angle, we would work backwards as before, using $\boxed{\text{INV}}$ $\boxed{\text{TAN}}$, $\boxed{\text{2ND FCT}}$ $\boxed{\text{TAN}}$, $\boxed{\text{ARC}}$ $\boxed{\text{TAN}}$, $\boxed{\text{TAN}^{-1}}$, etc.

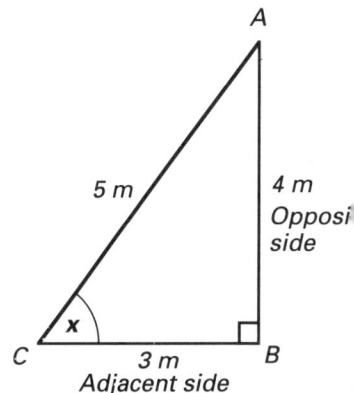

Example 8

tan 34.7° = 0.6924 (to 4 decimal places).
The angle whose tangent is 1.82 is ... 61.2° (to 1 decimal place).

EXERCISE 14.13

Find the tangents of the following angles, writing your answers to
four decimal places.

1 35°	**2** 78°	**3** 49°	**4** 87°	**5** 45°	**6** 7°
7 55°	**8** 45.1°	**9** 70.5°	**10** 5.7°	**11** 65.3°	**12** 32.6°
13 52.1°	**14** 7.8°	**15** 33.5°	**16** 80.5°	**17** 78.6°	**18** 34.9°
19 88.8°	**20** 51.7°	**21** 80.4°			

Find the angle, to one decimal place, whose tangent is:

22 0.6009	**23** 1.4181	**24** 2.8083	**25** 0.6669	**26** 0.4777	
27 0.1039	**28** 2.2497	**29** 1.0024	**30** 0.4550	**31** 3.5227	
32 1.3555	**33** 12.5	**34** 0.3928	**35** 0.9846	**36** 2.3085	
37 1.9643	**38** 0.9975				

Using tangent

Example 9

We will use the letter x for the side we want to find.

$$\frac{\text{opposite side}}{\text{adjacent side}} = \frac{BC}{AB} = \frac{x}{7}$$

$$\tan 30° = \frac{x}{7}$$

$$0.5774 = \frac{x}{7}$$

$$x = 0.5774 \times 7 = 4.0418 \text{ cm} = 4.04 \text{ cm (to 3 s.f.)}$$

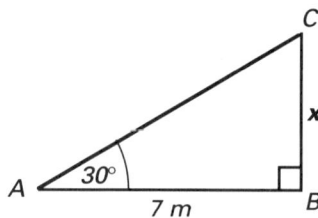

Example 10

We will use the letter x for the angle we want to find.

$$\frac{\text{opposite side}}{\text{adjacent side}} = \frac{ZY}{XY} = \frac{11.9}{8}$$

$$\tan x = \frac{11.9}{8}$$

$$\tan x = 1.4875$$

$$x = 56.1° \text{ (to 1 decimal place)}$$

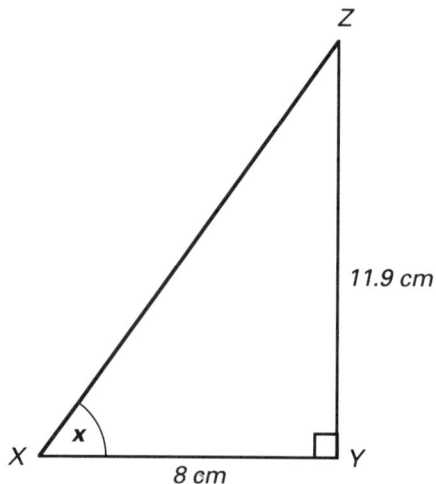

EXERCISE **14.14**

1

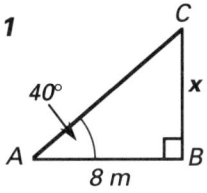

40°
8 m
A B
C
x

2

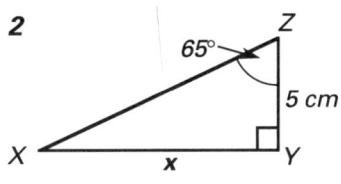

65°
Z
5 cm
X **x** Y

3 D

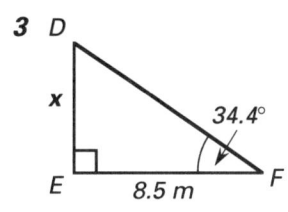

x
34.4°
E 8.5 m F

4 P

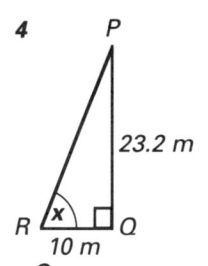

23.2 m
R **x** Q
10 m

5 B

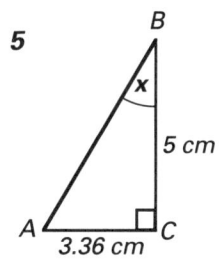

x
5 cm
A C
3.36 cm

6 S

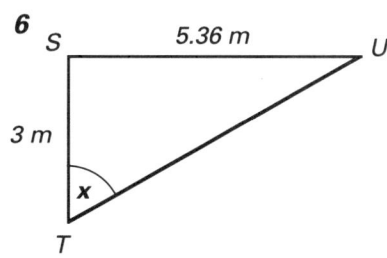

5.36 m U
3 m
x
T

7 D 4 m E

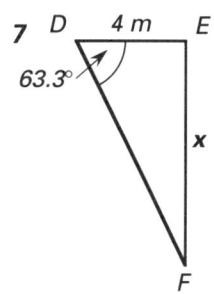

63.3°
x
F

8

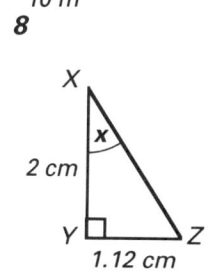

X
x
2 cm
Y Z
1.12 cm

9

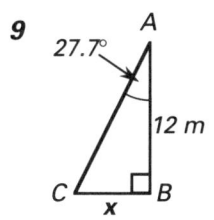

A
27.7°
12 m
C B
x

10 P

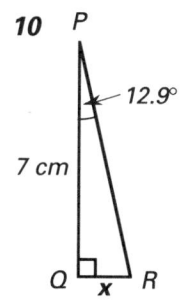

12.9°
7 cm
Q R
x

11

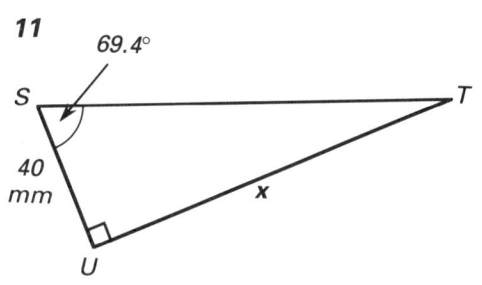

69.4°
S T
40 mm
x
U

12

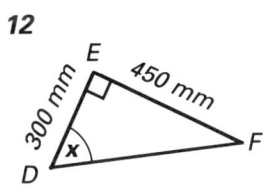

E
300 mm 450 mm
x F
D

13 B

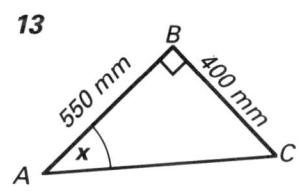

550 mm 400 mm
x C
A

14

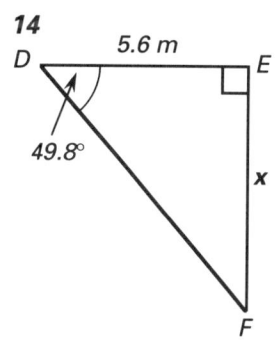

D 5.6 m E
49.8°
x
F

15

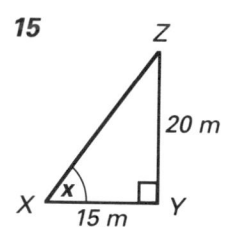

Z
20 m
X **x** Y
15 m

Investigation B

Investigate the tangent of angles. How does the decimal equivalent change as the angle gets bigger? What happens as the angle gets nearer and nearer to 90°? Also compare the tangent of small angles to the sine of the same small angles. What do you find?

Summary

$$\sin x = \frac{\text{opposite side}}{\text{hypotenuse}} = \frac{BC}{AC}$$

$$\cos x = \frac{\text{adjacent side}}{\text{hypotenuse}} = \frac{AB}{AC}$$

$$\tan x = \frac{\text{opposite side}}{\text{adjacent side}} = \frac{BC}{AB}$$

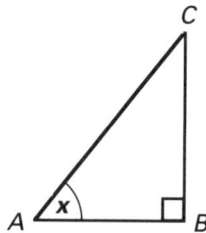

Many people like to remember the ratios as a single expression:

S O H C A H T O A

Each letter stands for an abbreviation, e.g. Sin, Opp, Hyp, Cos, Adj, Hyp, etc.

Example 11

Find x in the triangle.
First we have to find out which ratio to use. Look at the angle x. In the diagram we have both the *hypotenuse* and the *adjacent side*, which both belong to the *cosine* ratio, so this is the ratio we need to use.

$$\cos x = \frac{\text{adjacent side}}{\text{hypotenuse}} = \frac{2.55}{10} = 0.255$$

We want the angle whose cosine is 0.255, that is, 75.2°.

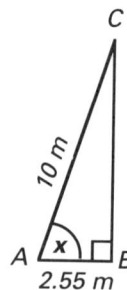

Example 12

Find x in the triangle.
In the diagram we have the *opposite side* and are trying to find the *adjacent side*, both of which are concerned with the *tangent* ratio, so this is the ratio we use.

$$\tan 40° = \frac{\text{opposite side}}{\text{adjacent side}} = \frac{5}{x}$$

The difficulty with this is that the x term is on the bottom of the fraction (the denominator), making the problem difficult to solve. We therefore go back to the diagram and use the other angle: $180° - 90° - 40° = 50°$.

$$\tan 50° = \frac{\text{opposite side}}{\text{adjacent side}} = \frac{x}{5}$$

$$1.1918 = \frac{x}{5}$$

$$x = 1.1918 \times 5 = 5.959 \text{ m} = 5.96 \text{ m (to 3 s.f.)}$$

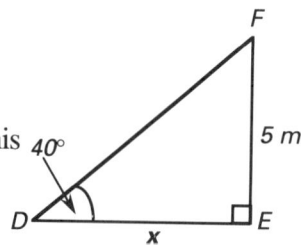

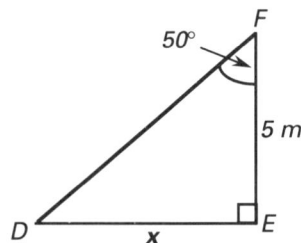

EXERCISE **14.15**

Find the missing side or angle marked as x in the following triangles.

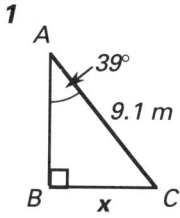

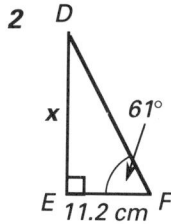

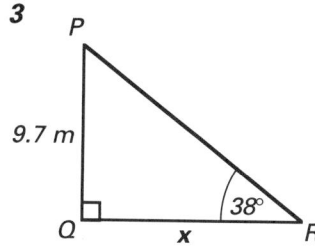

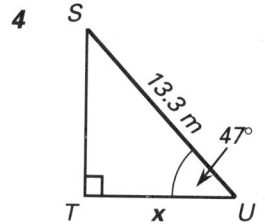

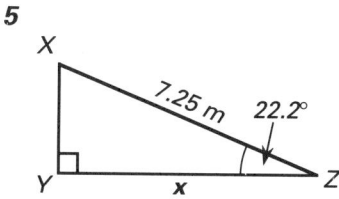

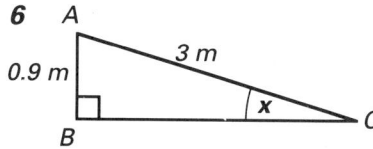

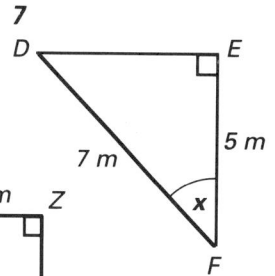

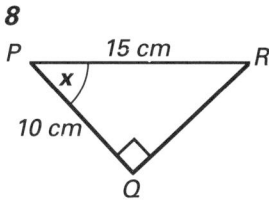

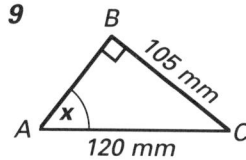

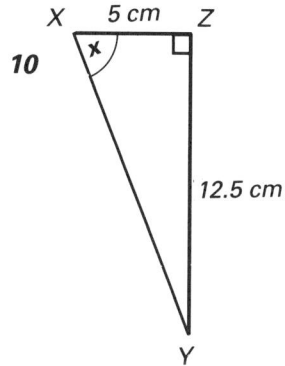

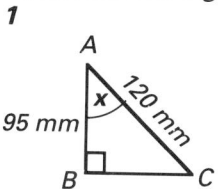

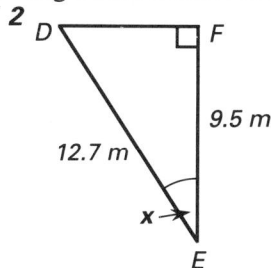

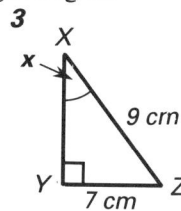

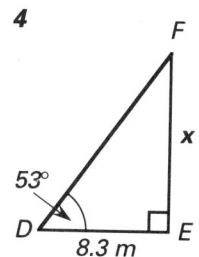

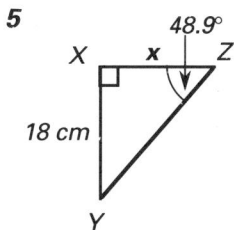

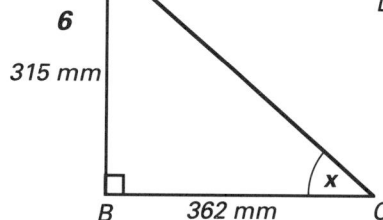

1

A 39°, 9.1 m, B, x, C

2

D, x, 61°, E 11.2 cm, F

3

P, 9.7 m, Q, x, 38°, R

4

S, 13.3 m, 47°, T, x, U

5

X, 7.25 m, 22.2°, Y, x, Z

6

A, 0.9 m, 3 m, B, x, C

7

D, E, 7 m, 5 m, x, F

8

P, 15 cm, R, x, 10 cm, Q

9

B, 105 mm, A, x, 120 mm, C

10

X, 5 cm, Z, x, 12.5 cm, Y

EXERCISE **14.16**

Find the missing side or angle marked as x in the following triangles.

1

A, x, 120 mm, 95 mm, B, C

2

D, F, 9.5 m, 12.7 m, x, E

3

X, x, 9 crn, Y, 7 cm, Z

4

F, x, 53°, D, 8.3 m, E

5

48.9°, X, x, Z, 18 cm, Y

6

A, 315 mm, B, 362 mm, x, C

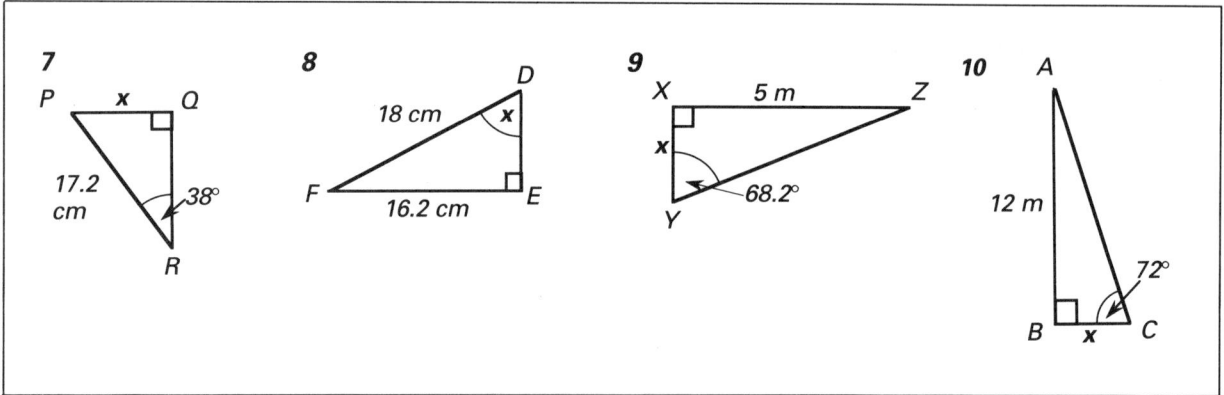

7

P — x — Q

17.2 cm   38°

R

8

18 cm   D   x

F   16.2 cm   E

9

X   5 m   Z

x   68.2°

Y

10

A

12 m

72°

B   x   C

Terminology

In using trigonometry we have to know the various ways of describing angles.

This angle could be described as (a) an angle below the horizontal, (b) an inclination below the horizontal, or (c) an angle of depression.

This angle could be described as (a) an angle above the horizontal, (b) an inclination above the horizontal, or (c) an angle of elevation.

This angle could be described as (a) an angle measured to the vertical, or (b) an angle inclined to the vertical.

EXERCISE 14.17

1 Find the angle at which the diagonal of the rectangle is inclined to one of the sides.

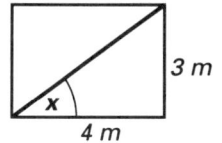

2 The angle of elevation of the top of a cliff from a boat 80 m away from its foot is 25°. Find the height of the cliff.

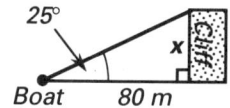

3 From a window 15 m up, the angle of depression of an object on the ground is 20°. Find the distance of the object from the base of the building.

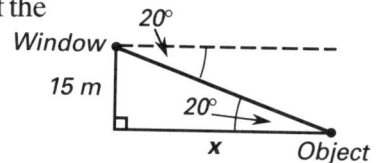

4 What is the height of a staircase 7.5 m long and inclined at 38° to the horizontal?

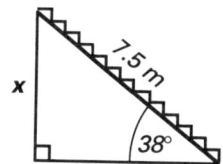

5 A ladder 10.6 m long leans against a vertical wall and makes an angle of 70° with the horizontal. How far is the foot of the ladder from the wall?

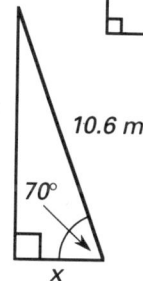

6 A tower 16 m high leans at an angle of 3.5° from the vertical. What is the perpendicular height of the tower?

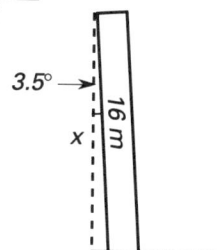

7 Find the height of the triangle.

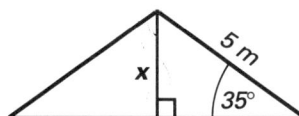

EXERCISE 14.18

1 Find *x* and *y*.

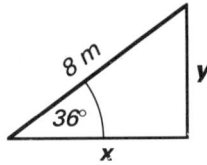

2 Find *x* and *y*.

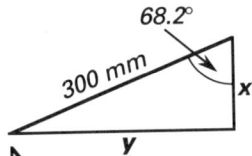

3 Find *x* and *y*.

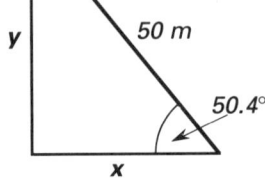

4 Find *x*.

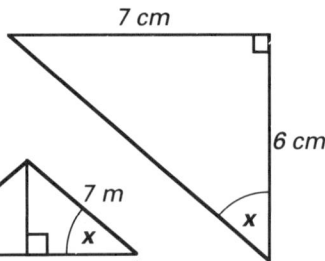

5 Find *x*.

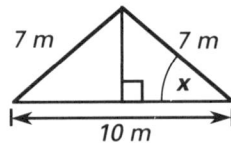

6 Find *x*.

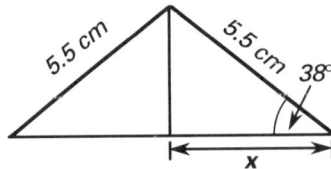

7 A bell tent has a diameter of 6 m and its pole is 4 m high. What angle does the slant side make with the ground?

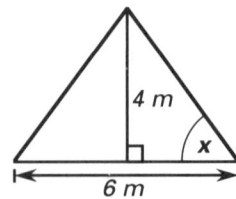

8 A pyramid structure has a base of length 7 metres and a slant height of 10 metres. Calculate (*a*) the angle *x* (*b*) the height of the pyramid.

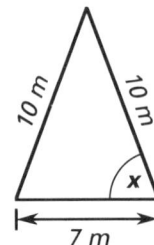

Investigation C

A useful practical exercise is to use trigonometrical theory to find the height of some of your school buildings. Surveyors use something called a theodolite to look along to measure the angle of elevation; you could make one yourself. Measure the distance from where you are standing to the base of the wall, and measure the angle. To find the actual height you will need to do some trigonometrical calculations. Remember to take into account the height at which you are holding your theodolite!

Revision exercises: Chapters 11 – 14

Find the missing number to replace the asterisk in each equivalent fraction:

1 $\dfrac{3}{5}=\dfrac{*}{10}$ **2** $\dfrac{2}{3}=\dfrac{*}{12}$ **3** $\dfrac{1}{6}=\dfrac{*}{30}$ **4** $\dfrac{7}{8}=\dfrac{*}{16}$ **5** $\dfrac{8}{10}=\dfrac{*}{5}$

6 $\dfrac{15}{20}=\dfrac{*}{4}$ **7** $\dfrac{2}{7}=\dfrac{*}{21}$ **8** $\dfrac{30}{36}=\dfrac{*}{6}$ **9** $\dfrac{7}{10}=\dfrac{*}{100}$ **10** $\dfrac{12}{50}=\dfrac{*}{25}$

Work out:

11 $\dfrac{2}{3}+\dfrac{1}{6}$ **12** $\dfrac{1}{4}+\dfrac{3}{8}$ **13** $\dfrac{1}{2}+\dfrac{1}{6}$ **14** $\dfrac{1}{6}+\dfrac{5}{12}$ **15** $\dfrac{5}{8}+\dfrac{1}{4}$

16 $\dfrac{1}{5}+\dfrac{1}{2}$ **17** $\dfrac{1}{3}+\dfrac{1}{4}$ **18** $\dfrac{1}{3}+\dfrac{2}{5}$ **19** $\dfrac{1}{7}+\dfrac{1}{2}$ **20** $\dfrac{1}{6}+\dfrac{1}{5}$

21 $\dfrac{3}{4}-\dfrac{1}{2}$ **22** $\dfrac{1}{2}-\dfrac{1}{3}$ **23** $\dfrac{2}{5}-\dfrac{1}{10}$ **24** $\dfrac{7}{8}-\dfrac{1}{4}$ **25** $\dfrac{9}{10}-\dfrac{1}{2}$

26 $\dfrac{5}{6}-\dfrac{1}{4}$ **27** $\dfrac{4}{5}-\dfrac{1}{2}$ **28** $\dfrac{3}{4}-\dfrac{7}{12}$ **29** $\dfrac{3}{5}-\dfrac{1}{4}$ **30** $\dfrac{17}{20}-\dfrac{1}{4}$

31 $4\dfrac{4}{5}-1\dfrac{1}{2}$ **32** $6\dfrac{1}{2}-2\dfrac{1}{4}$ **33** $7\dfrac{3}{4}+4\dfrac{1}{3}$ **34** $5\dfrac{3}{8}+1\dfrac{1}{6}$ **35** $2\dfrac{1}{2}+3\dfrac{1}{3}$

36 $7\dfrac{7}{8}-3\dfrac{5}{6}$ **37** $4\dfrac{4}{7}-1\dfrac{1}{4}$ **38** $3\dfrac{1}{4}+5\dfrac{3}{8}$ **39** $1\dfrac{1}{5}+2\dfrac{2}{3}$ **40** $5\dfrac{7}{8}-1\dfrac{1}{3}$

41 $4\dfrac{3}{4}+1\dfrac{1}{2}$ **42** $2\dfrac{3}{4}+\dfrac{7}{8}$ **43** $1\dfrac{4}{7}+1\dfrac{4}{7}$ **44** $5-3\dfrac{1}{3}$ **45** $6\dfrac{1}{2}-2\dfrac{3}{4}$

46 $2\dfrac{1}{5}-1\dfrac{1}{2}$ **47** $3\dfrac{3}{4}+2\dfrac{5}{6}$ **48** $4-1\dfrac{3}{5}$ **49** $6\dfrac{1}{6}-4\dfrac{1}{4}$ **50** $3\dfrac{2}{3}+3\dfrac{2}{3}+3\dfrac{2}{3}$

51 $\dfrac{1}{3}\times\dfrac{2}{5}$ **52** $\dfrac{3}{4}\times\dfrac{1}{2}$ **53** $\dfrac{2}{3}\times\dfrac{1}{4}$ **54** $\dfrac{1}{2}\times\dfrac{3}{7}$ **55** $\dfrac{2}{5}\times\dfrac{1}{6}$

56 $\dfrac{1}{7}\times\dfrac{4}{5}$ **57** $\dfrac{4}{5}\times\dfrac{1}{6}$ **58** $\dfrac{3}{8}\times\dfrac{2}{3}$ **59** $\dfrac{7}{8}\times\dfrac{4}{5}$ **60** $\dfrac{3}{8}\times\dfrac{2}{9}$

61 $2\dfrac{1}{2}\times1\dfrac{1}{2}$ **62** $1\dfrac{1}{4}\times\dfrac{1}{3}$ **63** $6\times1\dfrac{1}{3}$ **64** $3\dfrac{1}{8}\times\dfrac{4}{5}$ **65** $3\dfrac{1}{2}\times6$

66 $1\dfrac{3}{5}\times1\dfrac{3}{4}$ **67** $5\dfrac{1}{2}\times5\dfrac{1}{2}$ **68** $1\dfrac{2}{3}\times\dfrac{3}{5}$ **69** $11\dfrac{1}{9}\times\dfrac{7}{50}$ **70** $7\times2\dfrac{3}{4}$

71 $\dfrac{5}{6}\div\dfrac{2}{3}$ **72** $\dfrac{3}{8}\div\dfrac{1}{2}$ **73** $\dfrac{4}{5}\div\dfrac{1}{3}$ **74** $\dfrac{3}{4}\div\dfrac{2}{5}$ **75** $2\dfrac{3}{4}\div\dfrac{1}{2}$

76 $1\dfrac{1}{2}\div\dfrac{4}{7}$ **77** $\dfrac{3}{8}\div3$ **78** $3\div\dfrac{3}{8}$ **79** $2\div\dfrac{1}{4}$ **80** $3\dfrac{1}{2}\div1\dfrac{3}{4}$

Work out the value of these expressions:

81 £3.27 + £5.14 − £1.08

82 £20 − (£3.16 + £5.99)

83 (5 × 85p) + (3 × £1.15)

84 $\frac{3}{4}$ of £2.36

85 $\frac{7}{10}$ of £8.90 + $\frac{2}{3}$ of £8.31

86 $1\frac{2}{3}$ m of material costs £8.50. How much does 1 metre cost?

87 A 5 m length of cable costs £7.20. How much would it cost to buy
(a) 8 m (b) 8.75 m?

88 A piece of guttering $8\frac{1}{3}$ m long is bought at £1.20 per metre.
How much does this piece of guttering cost?

89 How many pieces of tubing $\frac{4}{5}$ m long can be cut from a length of 8 m
long?

90 A nail $\frac{7}{8}$ inch long is hammered through a piece of wood $\frac{5}{16}$ inch thick.
How far does the nail protrude through the piece of wood?

Find (a) the area (b) the perimeter of the following:

91

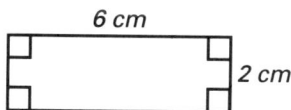

6 cm / 2 cm

92

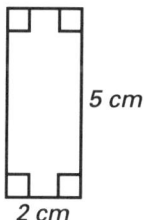

5 cm / 2 cm

93

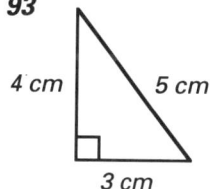

4 cm / 5 cm / 3 cm

94

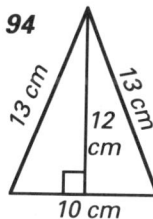

13 cm / 13 cm / 12 cm / 10 cm

95

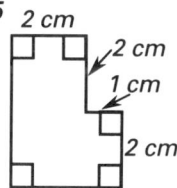

2 cm / 2 cm / 1 cm / 2 cm

96

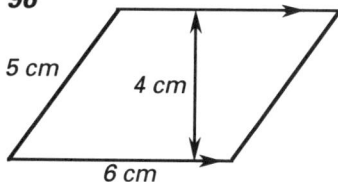

5 cm / 4 cm / 6 cm

97

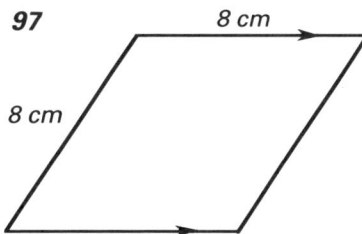

8 cm / 8 cm

98

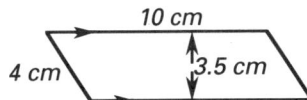

10 cm / 4 cm / 3.5 cm

Find the area of the shaded portion of each diagram:

99

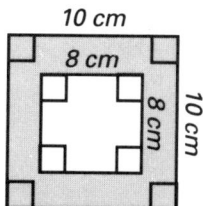

10 cm / 8 cm / 8 cm / 10 cm

100

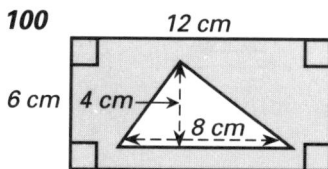

12 cm / 6 cm / 4 cm / 8 cm

101

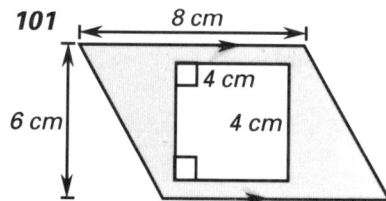

8 cm / 4 cm / 6 cm / 4 cm

Find the total area:

102

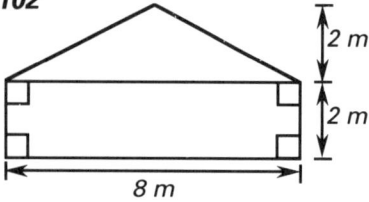

2 m

2 m

8 m

103

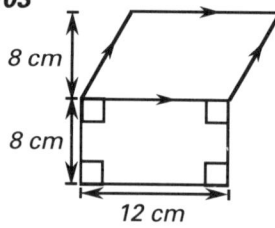

8 cm

8 cm

12 cm

104

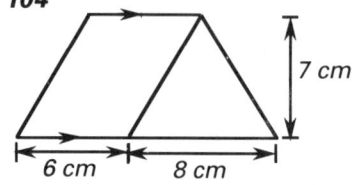

7 cm

6 cm   8 cm

Remember: circumference of a circle $= \pi \times$ diameter;
area $= \pi \times$ (radius)^2

Find (a) the circumference (b) the area of each of the following circles:

105 1 cm

106

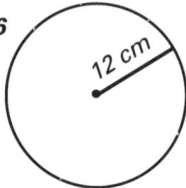

12 cm

107

4 cm

108 1.3 cm

109

10 m

110 4 m

111

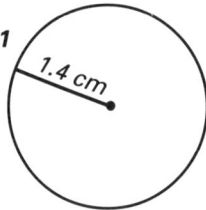

1.4 cm

112

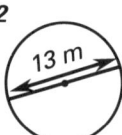

13 m

113

11.5 m

114

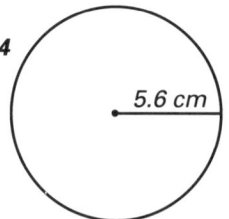

5.6 cm

For each of the following diagrams find the shaded area:

115

2 cm

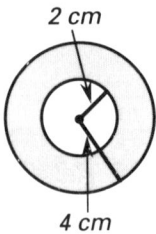

4 cm

116

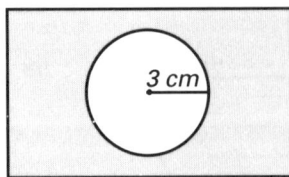

3 cm

117

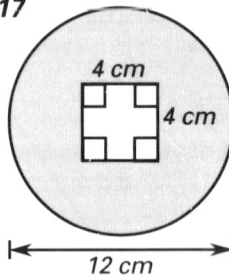

4 cm

4 cm

12 cm

118

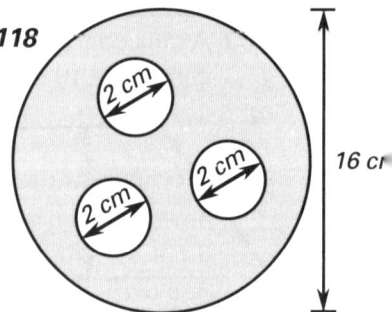

2 cm

2 cm

2 cm

2 cm

16 cm

Find (a) the area (b) the perimeter of the following:

119

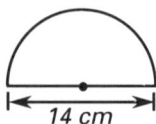

14 cm

120

8 cm

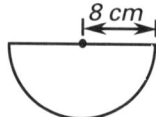

121

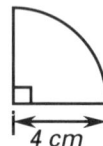

4 cm

122

3 cm

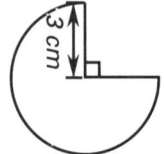

Find the volume of the following:

123

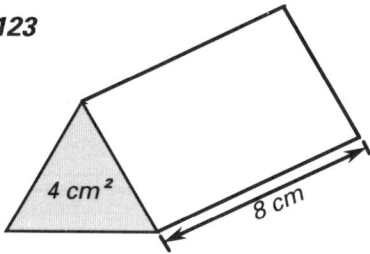

4 cm²   8 cm

124

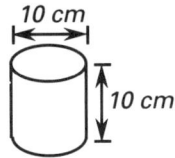

10 cm   10 cm

125

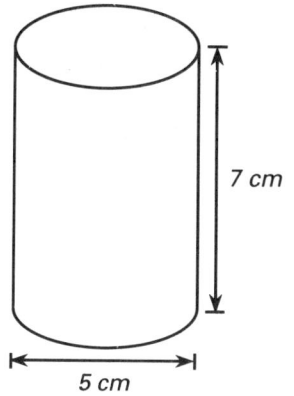

7 cm

126

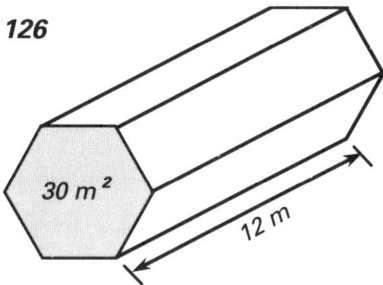

30 m²   12 m

127

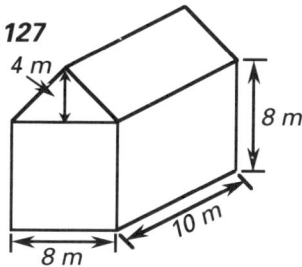

4 m   8 m   8 m   10 m

128

5 cm

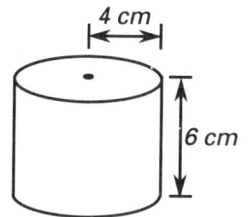

4 cm   6 cm

129

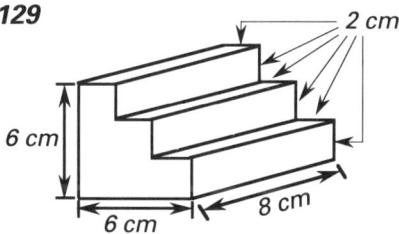

2 cm   6 cm   6 cm   8 cm

130

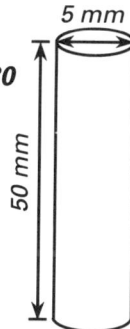

5 mm   50 mm

131

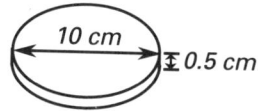

10 cm   0.5 cm

132

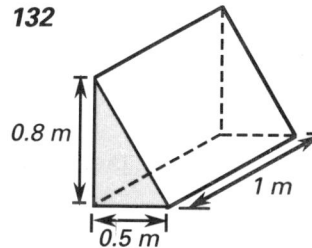

0.8 m   1 m   0.5 m

133 A tank containing 10 litres of liquid is to be emptied into cylindrical containers 10 cm high and 5 cm in diameter. How many containers will be needed? (1 litre = 1000 cm³.)

134 A cylindrical drum 3 m high and 0.8 m in diameter contains a powder which is to be transferred into packets of dimensions 0.1 m × 0.1 m × 0.2 m. How many packets will be filled?

135 The chocolate contents of the pan are to be poured into a machine that delivers the chocolate into triangular moulds, as shown in the diagram, which are on a conveyor belt. How many chocolates will be produced?

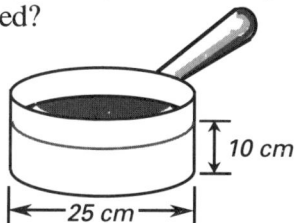

10 cm   25 cm

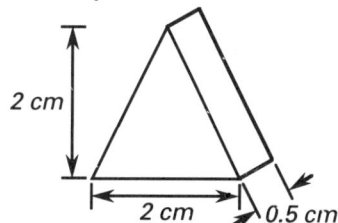

2 cm   2 cm   0.5 cm

136 A sack containing 50 000 cm³ of powder is to be emptied into containers of dimensions 10 cm × 10 cm × 30 cm. How many containers will be filled?

137 Draw a net for this solid.

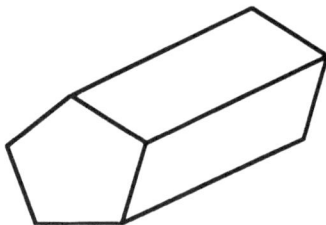

138 How many (*a*) edges (*b*) vertices will the solid have once it is made from the net?

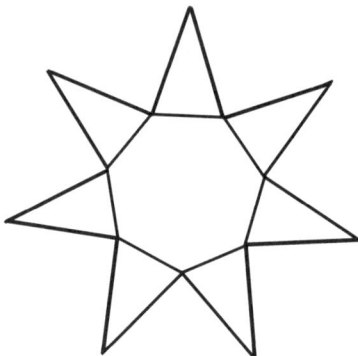

139 Draw a net for this solid.

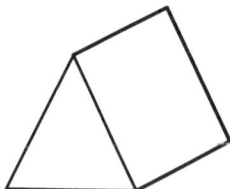

140 How many (*a*) edges (*b*) vertices will the solid have once it is made from the net?

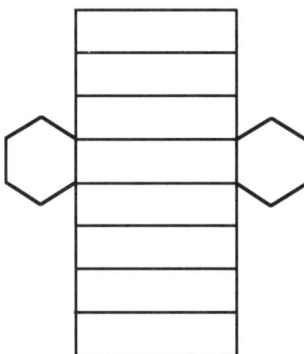

141 What shape will the net make when A, B and C are joined together?

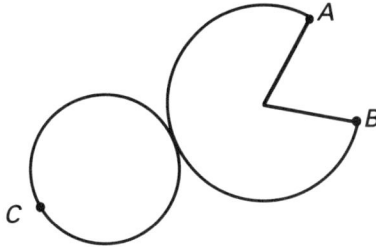

142 What shape will this net make once it is joined together?

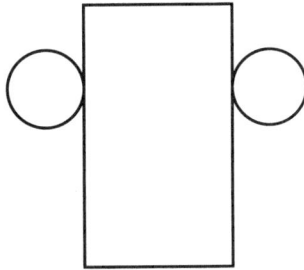

A card is selected at random from a pack of 52. Work out the probability that the card is:

143 a heart   **144** a black card   **145** a king

146 a seven   **147** a red three   **148** not a diamond

149 a picture card   **150** not a three or four   **151** the queen of spades

152 a heart or a king

A box contains four blue pens, three red pens, and two black pens. One is selected at random. Work out the probability that the pen is:

153 blue   **154** black   **155** not red

156 either black or red   **157** green   **158** not yellow

159 neither black nor blue   **160** either blue, black or red

I remove the three red pens, leaving the four blue and two black. I now pick one, keep it and then pick another. Work out the probability that the pens that have been picked are:

161 two blue   **162** two black   **163** one of each

164 both the same colour

Two dice are thrown, and the numbers added up. Copy and complete the table.

	1	2	3	4	5	6
1	2	3	4	5	6	7
2	3	.	.	.	.	.
3	.	.	.	7	.	.
4	.	.	.	.	.	.
5	.	.	.	.	10	.
6	.	8	.	.	.	12

Work out the probabilities of throwing:

165 12

166 7

167 3 or less

168 8 or 9

169 an even total

170 at least one 6

171 13

Use your calculator to write down the value given by the following, correct to four decimal places.

172 $\sin 20°$ **173** $\cos 48°$ **174** $\tan 62°$ **175** $\cos 34°$ **176** $\tan 40°$

177 $\sin 41°$ **178** $\cos 28°$ **179** $\tan 71°$ **180** $\sin 82°$ **181** $\cos 63°$

182 $\tan 67°$ **183** $\sin 35.1°$ **184** $\tan 75.2°$ **185** $\cos 35.8°$ **186** $\sin 74.4°$

187 $\sin 32.7°$ **188** $\cos 50.2°$ **189** $\tan 83.5°$ **190** $\sin 61.9°$ **191** $\tan 24.8°$

Find, correct to one decimal place, the angle whose sine is:

192 0.3127 **193** 0.4052 **194** 0.9814 **195** 0.0423 **196** 0.6812

Find, correct to one decimal place, the angle whose cosine is:

197 0.0321 **198** 0.5221 **199** 0.7272 **200** 0.9104 **201** 0.1097

Find, correct to one decimal place, the angle whose tangent is:

202 0.3328 **203** 1.4137 **204** 0.9841 **205** 2.5422 **206** 0.6002

Find the length x in these triangles:

207

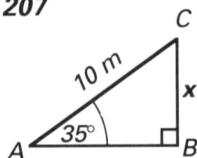

208

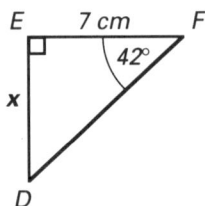

209

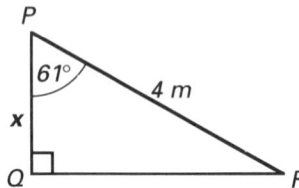

210

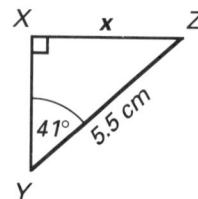

211

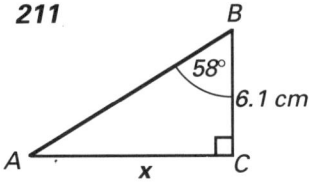

58°
6.1 cm
x
B
A
C

212

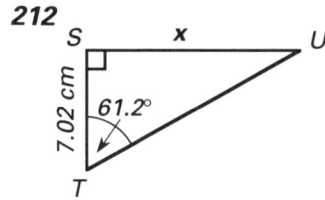

S
x
U
7.02 cm
61.2°
T

213

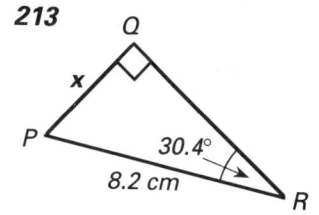

Q
x
P
30.4°
8.2 cm
R

214

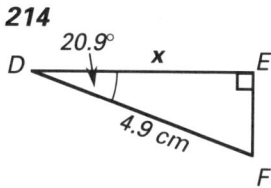

20.9°
x
D
E
4.9 cm
F

215

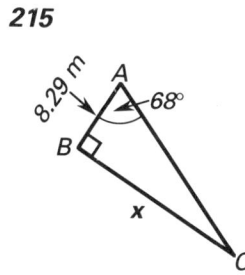

8.29 m
A
68°
B
x
C

216

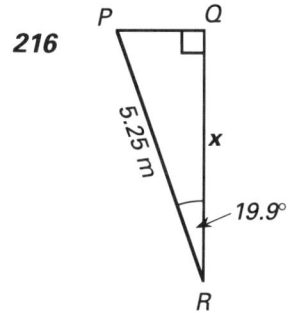

P
Q
5.25 m
x
19.9°
R

Find the angle *x* in these triangles:

217

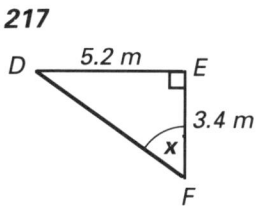

5.2 m
D
E
3.4 m
x
F

218

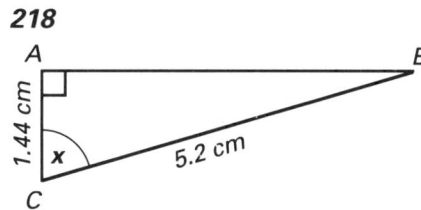

A
B
1.44 cm
x
5.2 cm
C

219

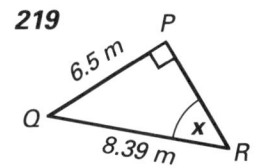

P
6.5 m
Q
x
8.39 m
R

220

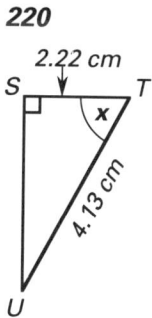

2.22 cm
S
T
x
4.13 cm
U

221

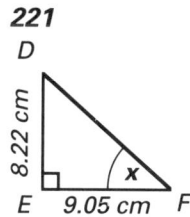

D
8.22 cm
x
E
9.05 cm
F

222

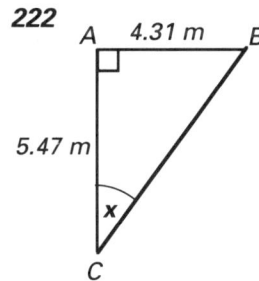

A
4.31 m
B
5.47 m
x
C

223

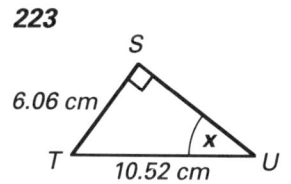

S
6.06 cm
x
T
10.52 cm
U

224

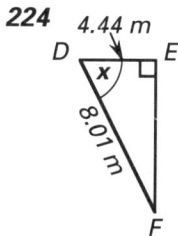

4.44 m
D
E
x
8.01 m
F

225

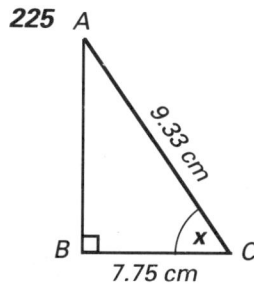

A
9.33 cm
B
x
C
7.75 cm

226

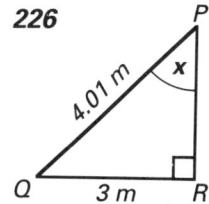

P
4.01 m
x
Q
3 m
R

Find *x* in these triangles:

227

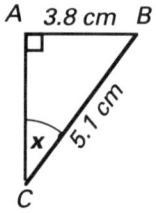

228

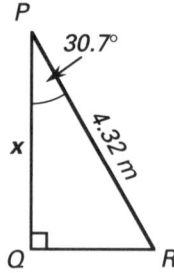

229

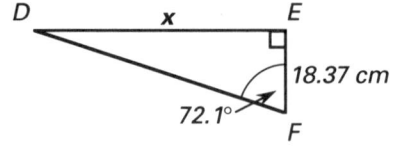

230

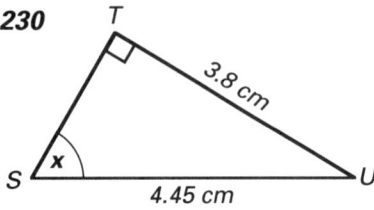

231

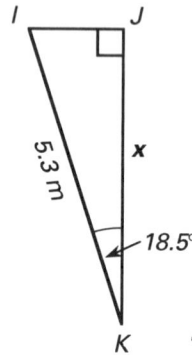

232

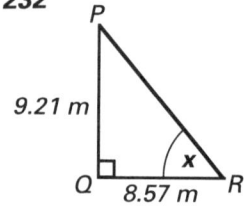

233

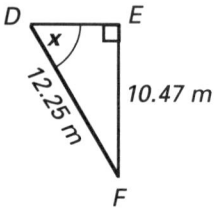

234

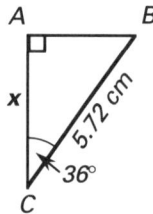

235

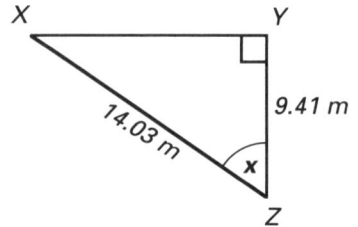

236

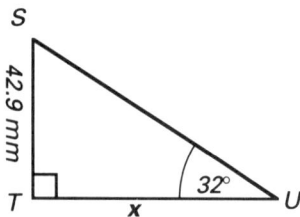

237

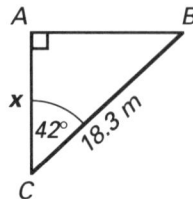

238

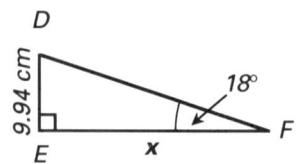

239

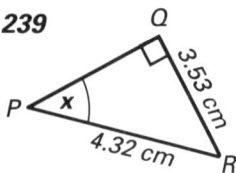

240

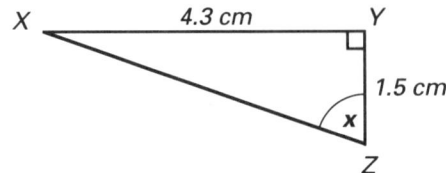

241

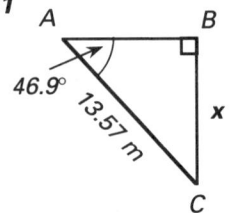

242 A towel is to have a diagonal strip of material added to it. At what angle, x, should it be sewn?

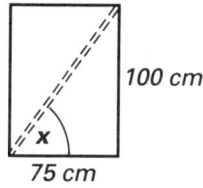

100 cm

x

75 cm

Julie's garden slopes at an angle of 17° to the horizontal. If the string she has used is 14 metres long, how far does the garden fall vertically?

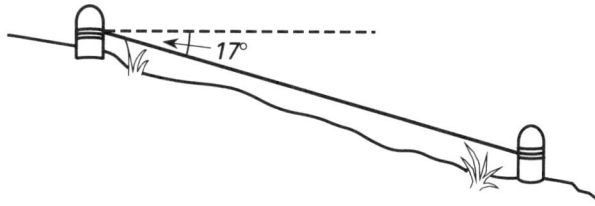

17°

244 A hole is to be drilled through the outside wall of a house as shown. At what angle should the drill be inclined to the horizontal?

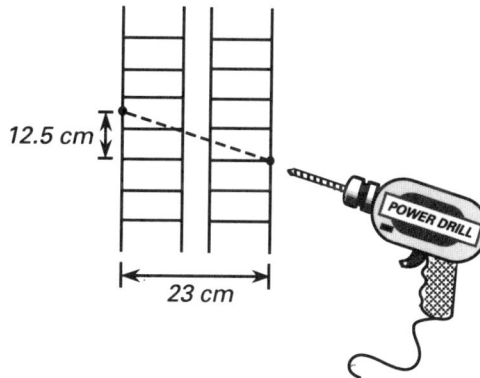

12.5 cm

POWER DRILL

23 cm

245 A section of guttering slopes downwards to allow for water drainage. At what angle to the horizontal does the gutter slope?

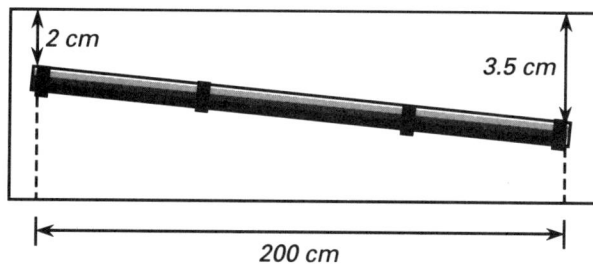

2 cm

3.5 cm

200 cm